Stamatios Papadakis and Georgios Lampropoulos (Eds.)

Intelligent Educational Robots

Also of interest

Semantic Intelligent Computing and Applications
De Gruyter Frontiers in Computational Intelligence, Vol. 16
Mangesh M. Ghonge, Pradeep Nijalingappa, Renjith V. Ravi,
Shilpa Laddha, Pallavi Vijay Chavan (Eds.), 2023
ISBN 978-3-11-078159-5, e-ISBN (PDF) 978-3-11-078166-3,
e-ISBN (EPUB) 978-3-11-078174-8

Personalized Human-Computer Interaction
Mirjam Augstein, Eelco Herder, Wolfgang Wörndl (Eds.), 2023
ISBN 978-3-11-099960-0, e-ISBN (PDF) 978-3-11-098856-7,
e-ISBN (EPUB) 978-3-11-098877-2

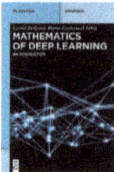

Mathematics of Deep Learning.
An Introduction
Leonid Berlyand, Pierre-Emmanuel Jabin, 2023
ISBN 978-3-11-102431-8, e-ISBN (PDF) 978-3-11-102555-1,
e-ISBN (EPUB) 978-3-11-102580-3

Digital Transformation in Healthcare 5.0.
Volume 1: IoT, AI and Digital Twin
Rishabha Malviya, Sonali Sundram, Rajesh Kumar Dhanaraj,
Seifedine Kadry, 2024
ISBN 978-3-11-132646-7, e-ISBN (PDF) 978-3-11-132785-3,
e-ISBN (EPUB) 978-3-11-132786-0

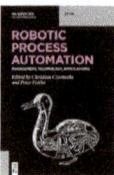

Robotic Process Automation.
Management, Technology, Applications
Christian Czarnecki, Peter Fettke (Eds.), 2021
ISBN 978-3-11-067668-6, e-ISBN (PDF) 978-3-11-067669-3,
e-ISBN (EPUB) 978-3-11-067677-8

Intelligent Educational Robots

Toward Personalized Learning Environments

Edited by
Stamatios Papadakis and Georgios Lampropoulos

DE GRUYTER

Editors

Stamatios Papadakis
Assistant Professor
Department of Preschool Education
School of Education
University of Crete
Greece
stpapadakis@uoc.gr

Georgios Lampropoulos
Postdoctoral Researcher
Department of Applied Informatics
School of Information Sciences
University of Macedonia
Greece
And
Visiting Lecturer
Department of Education
School of Education
University of Nicosia
Cyprus
And
Postdoctoral Researcher
Department of Preschool Education
School of Education
University of Crete
Greece
glampropoulos@uom.edu.gr

ISBN 978-3-11-135206-0
e-ISBN (PDF) 978-3-11-135269-5
e-ISBN (EPUB) 978-3-11-135296-1

Library of Congress Control Number: 2024944872

Bibliographic information published by the Deutsche Nationalbibliothek
The Deutsche Nationalbibliothek lists this publication in the Deutsche Nationalbibliografie;
detailed bibliographic data are available on the internet at http://dnb.dnb.de.

© 2025 Walter de Gruyter GmbH, Berlin/Boston
Cover image: D-Keine/E+/Getty Images
Typesetting: Integra Software Services Pvt. Ltd.

www.degruyter.com

Preface

This book focuses on recent advances in maker education and on the integration of artificial intelligence (AI) and intelligent educational robots (IER) in P-12 education. Specifically, it covers various topics and trends regarding the advancements of maker education and the use of IER, AI, intelligent agents, and machine learning in P-12 education. It promotes students' motivation, critical thinking, and problem-solving skills through adaptive, interactive, and personalized learning experiences that capitalize on digital pedagogies, human-robot interaction, IER, and social robots.

The main aim of the book is to offer an overview of recent research into the adoption, integration, advancements, and impact of IER on education in the context of maker education. The topics include the application of various technologies and educational approaches for realizing AI, IER, and enhancing maker education. It presents findings and discussions on applications, addresses open issues and challenges, offers solutions, and provides suggestions for future lines of research for achieving AI and IER in education. This book assists researchers, practitioners, professionals, and academicians of various scientific domains in exploring and better comprehending the state of the art of maker education, AI and IER, their advancements, their impact, and future potentials and benefits in education.

This book contains chapters focusing on different pedagogical scaffolds to assist teachers in adopting and integrating intelligent robots and maker education in classrooms in P-12 education. The chapters include research-driven case studies, overviews, and standards-aligned lesson plans from real-life settings.

In Chapter 1, Lampropoulos and Papadakis examine the use of AI and educational robotics in maker education. Specifically, the chapter provides an overview of the existing literature regarding the current state of maker education. Additionally, it details the use of AI in education and its role in maker education. The chapter also delves into the use of educational robotics and how they can be adopted in teaching and learning activities in maker education. Furthermore, it goes into more detail regarding the benefits that can be yielded for teachers and students through this combination and offers conclusive remarks and suggestions for future research directions.

In Chapter 2, Zviel-Girshin and Nathan Rosenberg explore the transformative journey from making skills to engineering skills, facilitated by educational robotics playgrounds designed for learners of all ages. Drawing inspiration from educational theorists like Montessori, Dewey, Vygotsky, and Papert, the chapter outlines a project-based learning environment where hands-on activities are paramount. This approach is crystallized in the concept of "magineering," a blend of making, engineering, and manufacturing, which serves as a foundation for a creative and innovative educational ecosystem. The chapter highlights the significance of these playgrounds, known as engineering robotic playgrounds, in nurturing a seamless transition from simple-making activities to complex engineering tasks, thereby fostering a generation of innovators equipped with both practical skills and theoretical knowledge.

https://doi.org/10.1515/9783111352695-202

In Chapter 3, Carius and Baldner explore the implementation of the new high school curriculum in Brazil, focusing on integrating computational robotics, programming, and maker activities across the country's 26 states. The study evaluates the influence of the maker movement on STEM education, emphasizing the development of computational thinking in students. It combines qualitative analysis of state resolutions and quantitative data from the 2023 School Census to assess the infrastructure and adoption of educational robotics. Despite innovative initiatives, the chapter concludes that significant investment in school infrastructure is still needed to fully embrace technology and innovation in Brazilian education.

In Chapter 4, López-Bouzas et al. review the integration of educational robotics and computational language in school settings, highlighting their potential to enhance digital competencies among teachers and students. It discusses the importance of computational thinking, which includes skills like algorithmic thinking and problem-solving, which are essential for interacting with advanced technologies. The chapter emphasizes the need for teacher training to effectively incorporate educational robotics into the curriculum, addressing both the benefits and challenges, such as resistance due to lack of training and resources. It also explores the impact of educational robots on students' communicative, socio-emotional, and computational skills, presenting various studies and methodologies used to assess these outcomes. The chapter concludes by stressing the importance of systematic methodologies and larger sample sizes in future research to better understand robotics's educational benefits.

In Chapter 5, Conti et al. explore the integration of psychological measurements in educational robotics research. It highlights the increasing use of educational robots in classrooms to enhance various skills among K-12 students, such as social, communication, and cognitive abilities. The chapter emphasizes the importance of using standardized psychological tests to evaluate the effectiveness of these robots, as current research often lacks quantitative assessments. It reviews various psychological measurement tools like surveys, self-reports, and standardized questionnaires and discusses their application in assessing human-robot interactions. The chapter concludes by stressing the need for reliable and valid measurement tools to better understand and improve the interactions between students and educational robots.

In Chapter 6, Cersosimo and Pennazio discuss a research project to enhance social communication skills in children and adolescents with autism spectrum disorder (ASD) through social robots and virtual environments. The project, funded by the Italian Foundation for Autism, builds on previous studies that have shown the effectiveness of social robots in opening communicative channels for individuals with autism. The chapter outlines the project's methodology, which includes individualized work sessions with robots like Buddy, Navel, and NAO, designed to develop specific social skills. These sessions are conducted in rehabilitation and school contexts to ensure that learned skills are generalizable. The chapter also goes over a research and training laboratory for teachers, which helps equip them with the skills to design effective learning paths for autistic students using robotic tools.

In Chapter 7, Trixler and Pusztafalvi explore the potential of robots in autism-specific education, highlighting both opportunities and barriers. The chapter discusses how robots can create a personalized and predictable learning environment, which can help develop social and cognitive skills in individuals with autism. The chapter emphasizes the importance of careful pedagogical planning and the need for more in-depth research to evaluate long-term effects. It also addresses the challenges professionals face in integrating these technologies into educational settings, including the need for specialized training and resources. Overall, the chapter underscores robots' promising yet complex role in enhancing the quality of life for people with autism.

In Chapter 8, Álvarez-Herrero explores the integration of IER in primary education, highlighting their potential to enhance critical thinking and problem-solving skills. It discusses the various pedagogical benefits of IER, such as fostering creativity, collaboration, and hands-on learning. The chapter also presents a series of six activities designed to utilize IER effectively in the classroom, each with clear objectives, required materials, and assessment criteria. Additionally, it addresses the challenges of implementing IER, including teacher training, curriculum alignment, and ethical considerations, emphasizing the need for careful planning and support to maximize their educational impact.

In Chapter 9, Baena-Luna and García-Río explore the innovative use of intelligent educational robots in learning environments, highlighting their potential to transform traditional educational approaches. The chapter identifies key research trends, influential authors, and significant publications in this field through a bibliometric analysis of 169 selected papers from the Web of Science and Scopus databases. The analysis reveals the evolution of educational robotics, emphasizing their role in enhancing problem-solving skills, critical thinking, and collaboration among students. Despite the benefits, challenges such as the need for teacher training and the adaptation of robots to individual learning styles are noted. The chapter concludes by discussing future research opportunities and the integration of AI in educational robots.

In Chapter 10, Dinler explores the transformative potential of IER in early childhood education. It highlights how IERs can foster creativity, personalize learning, and promote inclusivity. The chapter provides a comprehensive list of commercially available IERs, emphasizing their diverse functionalities and age appropriateness. It underscores the importance of integrating IER to create a holistic and inclusive learning experience, incorporating STEM and maker culture into the curriculum. The chapter also addresses challenges such as equitable access to resources and the need for sophisticated software development. Ultimately, it envisions a future where IER are crucial in transforming early childhood education and providing equitable access to quality education for all children.

In Chapter 11, Stenová et al. investigate the interactivity of smart educational robots and their role in the educational process. The chapter categorizes interactivity into input, output, and input-output. Input interactivity involves sensors that gather environmental data, while output interactivity includes the robot's responses, like lights and sounds. Input-output interactivity combines both, with robots responding automatically to inputs. The chapter highlights the significance of these interactive features in developing twenty-first-century skills and STEAM education, enhancing student motivation and engagement. It also discusses the broader applications of educational robotics, such as language learning and support for autistic children, emphasizing the importance of integrating these technologies into the curriculum to prepare students for future challenges.

In Chapter 12, Luria explores the integration of AI chatbots in English language learning among higher education students in Israel. The study employs a mixed-method design, combining qualitative insights from semi-structured interviews and quantitative data from validated questionnaires. Key findings reveal that chatbot interactions significantly improve language learning outcomes, including performance, efficiency, motivation, and self-confidence. The thematic analysis highlights chatbots' perceived effectiveness, convenience, engagement, and confidence-building. Quantitative analyses show substantial improvements in language skills, supported by large effect sizes and positive correlations. The chapter highlights that AI chatbots can constitute practical tools for enhancing language acquisition in higher education, offering personalized, interactive, and accessible learning experiences.

In Chapter 13, Sharonova and Avdeeva explore the ethical models of teachers in the context of the fourth technological revolution and the integration of AI in education. The chapter examines the ethical challenges posed by AI, such as embedding ethical standards in algorithms and the interaction between humans and robots. The chapter contrasts traditional teacher-centered and student-centered models with AI-driven models, highlighting the unique qualities of human teachers – charisma, intuition, goal-setting, social imagination, and identity – that AI cannot replicate. The chapter highlights that although AI can assist in routine educational tasks, the irreplaceable human elements of teaching are crucial for the holistic development of students.

In Chapter 14, Papasarantou et al. present the "DIY robotic car controlled using voice commands" project, developed under the AI4STEM Erasmus + project. The project is designed for students aged 12 to 16 and introduces AI and IoT through the 5 Big Ideas of AI, project-based learning, and robotics. It consists of five activities, each accompanied by educational materials such as teacher guidelines and student worksheets. The project has been tested in workshops and junior high schools and utilizes the BBC micro:bit microcontroller and block-based programming to make implementation easier.

Contents

Georgios Lampropoulos and Stamatios Papadakis

Maker education and educational robotics in the age of artificial intelligence

Abstract: This chapter explores the influence of educational robotics and artificial intelligence in maker education. According to the findings, integrating artificial intelligence and educational robotics in maker education can enrich the educational process, create collaborative learning environments, and provide benefits for both teachers and students. In such cases, students can also benefit from personalized learning opportunities and acquire a deeper understanding through interactive and experiential learning activities. By involving students in such maker activities, their learning outcomes can be increased, their learning engagement, focus, and motivation can be enhanced, and their critical thinking and creativity can be improved. Additionally, students can improve their soft skills, digital skills, and artificial intelligence literacy.

Keywords: Artificial intelligence, AI, educational robotics, educational robots, maker education, maker movement, education

1 Introduction

Maker education is becoming more widely used across educational levels [1]. Makers are the individuals who are enthusiastic about and keen on developing, collaborating, and sharing [2]. Hence, it comes as no surprise that the maker movement puts emphasis on the needs of an individual to be actively involved, interact, and communicate within a community of shared interests and values [2, 3]. As a result, the maker movement utilizes information and communication technologies (ICTs) to promote and facilitate community building, interactions, communication, and creativity [4, 5].

Furthermore, maker education involves makers, makerspaces, and the act of making and engages learners in the creative development of artifacts and their distribution within the community using physical and digital means [4, 5]. In this context, learners participate in problem-based activities that foster their soft skills and creativity [6, 7] as they enable learners to create, experiment, and explore [7]. Additionally, through maker activities, a culture of learning by doing is promoted in which learners

Corresponding author: Georgios Lampropoulos, Department of Applied Informatics, School of Information Sciences, University of Macedonia, Thessaloniki, Greece; Department of Education, School of Education, University of Nicosia, Nicosia, Cyprus; Department of Preschool Education, School of Education, University of Crete, Greece, e-mail: glampropoulos@uom.edu.gr

Stamatios Papadakis, Department of Preschool Education, School of Education, University of Crete, Greece, e-mail: stpapadakis@uoc.gr

https://doi.org/10.1515/9783111352695-001

can develop socially and cognitively and enhance their innovation ability [8, 9]. However, there are several aspects that should be considered to effectively introduce and integrate these activities in classrooms with students' needs and characteristics playing a vital role in their overall learning engagement and motivation in maker activities [3, 6].

Given the close relationship between ICT and maker education, the rapid ICT advances can enrich maker education and increase its benefits. Simultaneously, students are looking for more personalized and experiential learning experiences [10], which can be achieved through the use of artificial intelligence [11] and educational robotics [12]. Moreover, the adoption and use of project-based learning and problem-based learning are gaining ground in the educational domain due to the benefits they can yield [13, 14]. These approaches can be further enriched when used in combination with artificial intelligence and educational robotics. Although maker education is more widely being used in teaching and learning activities to promote students' engagement, interest, and motivation and enable them to play an active role in shaping their own learning [2], not much is known about how artificial intelligence can support and enrich maker education. Therefore, the aim of this chapter is to examine the role of artificial intelligence and educational robots in maker education. The remainder of the chapter is structured as follows: In Section 2, the role of artificial intelligence in teaching and learning activities and in maker education is showcased. In Section 3, the role of educational robotics in maker education is presented. Finally, in Section 4, the outcomes are further discussed, conclusive remarks are provided, and future research directions are given.

2 Artificial intelligence in education and maker activities

Artificial intelligence can support maker education throughout the design, development, implementation, and sharing phases. Additionally, artificial intelligence can be used to support both teachers and students, enrich educational activities, and enhance educational outcomes [11, 15, 16]. Hence, research into the integration of artificial intelligence in education is increasing [17]. Artificial intelligence systems are characterized by being autonomous, adaptable, perceptive, rational, and able to learn and reason [18–20]. These systems can identify, process, interpret, and learn from data, simulate the way humans learn, interact, and reason, and can make autonomous and rational decisions [20–22].

Although there are numerous benefits that can be gained when introducing artificial intelligence into classrooms, there still remain several challenges and theoretical gaps that need to be addressed [23, 24]. When integrated in a student-centered manner, artificial intelligence can promote self-regulated learning, enhance personalized learning, and enable learners to be actively involved and cocreate their own learning in both formal and informal environments [25–27]. Moreover, artificial intelligence can be used in online, face-to-face, and blended learning environments [28] to offer adaptive and personalized learning opportunities [29, 30].

In the context of maker education, artificial intelligence can offer autonomous and rational data-driven recommendations and decision-making and improve personalized learning through individualized feedback. It can also aid teachers in providing guidance and instructions that are tailored to each individual learner's requirements and needs. Additionally, it can help automate some of the teachers' specific, routine tasks, which, in turn, enables them to pay closer attention to their students and dedicate more time to them. Artificial intelligence can also increase access to educational resources and improve the design of effective maker activities. Through recommender systems, both teachers and students can identify suitable material for their needs. By analyzing students' data, artificial intelligence systems can provide learners with personalized feedback tailored to their needs and skill activities and resources, thus improving self-regulated learning, which in turn enables learners to get a deeper understanding of the subject taught and promotes their development of a growth mindset. Using artificial intelligence, opportunities to create artificial intelligence-based projects to be explored in a hands-on manner are created. Hence, artificial intelligence can aid students in improving their technical skills and digital literacy. Therefore, artificial intelligence can support maker education, aid both students and teachers, and help create meaningful and engaging maker activities in which students play an active role. However, to effectively adopt and integrate artificial intelligence within maker education, there is a need to further examine and improve education stakeholders' artificial intelligence literacy.

3 The role of educational robotics in maker education

As an interdisciplinary field of study, educational technologies have roots in various domains and put emphasis on integrating robotics that were specifically designed and created for educational purposes in teaching and learning activities to develop, apply, and evaluate interactive and meaningful pedagogical activities and learning experiences [31, 32]. Due to the potential of educational robotics to enrich and transform the educational process and address students' educational requirements [33], the research surrounding their application in the educational domain has been increasing [34].

There are several benefits that can be yielded when introducing educational robotics in classrooms [35, 36]. Although different in nature and degree, positive students' learning outcomes have been observed when integrating robotic-based activities in the curriculum of all educational levels [37]. Besides the benefits that educational robotics can bring about for students, teachers favorably assess the use of educational robotics in teaching and learning activities [38]. Nonetheless, there is a need to create and provide teachers with appropriate training and development programs to cultivate their digital skills, improve their artificial intelligence literacy, and assist them in efficiently using adopting and using educational robotics in their classrooms [39].

Since there are different educational robotic types that can affect the educational activities differently [40], it is important to take students' traits, skills, knowledge, and characteristics into account when selecting educational robotic-based activities to select the most appropriate resources [41]. Educational robotics can also be enriched with artificial intelligence capabilities to offer personalized and adaptive learning [42]. In this context, through intelligent educational robotics, students can receive individualized guidance and feedback which, in turn, can improve their learning outcomes [43]. Through such activities, active and engaging learning experiences can be created [44], which promote social interactions, inclusive and collaborative learning [45–47], and enhance students' soft skills, learning motivation, engagement, and outcomes [48–50]. Additionally, by engaging in educational robotics-based learning activities, students' creativity, critical thinking, and computational thinking can be improved [51, 52].

Owing to their multidisciplinary and versatile nature as well as their ability to offer engaging experiential learning opportunities, educational robotics are well-versed to use within maker activities. Learning using educational robotic projects can take place in both informal and formal learning environments and enhance students' soft and digital skills through the combination of aspects from different disciplines. Furthermore, using educational robotics in the context of maker education can result in the creation of interactive and collaborative learning experiences that foster social interactions and communication. Due to the tangible nature of educational robotics and the iterative design process that characterizes robotics-based learning activities, students can engage in meaningful hands-on experiences that foster a growth mindset and enable a deeper understanding of the subjects taught. When using artificial intelligence and educational robotics within maker education, students can learn in a personalized way, and at their own pace, experience differentiated instructions and experiential learning, and pursue activities and projects that they are passionate about and interested in.

4 Conclusions

Maker education focuses on learners' active participation and engagement in teaching and learning activities and aims at enabling learners to become cocreators and codesigners of their own learning. Through the project-based learning activities it offers, maker education can support experiential learning and promote collaborative learning. Within maker activities, learners' soft skills, reasoning, and digital literacy can be improved. Educational robotics and artificial intelligence can be used in the context of maker education to further amplify its positive learning outcomes. Within such environments, learners engage in meaningful hands-on project-based activities that can increase their problem-solving skills, critical thinking, and creativity, as well as their higher-order thinking and social skills. Additionally, adaptive learning environments that focus on providing learners with personalized learning experiences can be created. Such environments promote learners' self-regulated learning and through data-driven decision-making and recommendations can provide them with individualized support, evaluation, and feedback. This combination can support both learners and educators and facilitate their daily tasks, so learners can focus on engaging in meaningful learning and educators can focus on supporting their students. By adopting artificial intelligence, educational robotics, and maker education, teaching and learning activities can be enriched, and various educational benefits, such as increased learning engagement, motivation, and achievements, can be achieved.

Future research should focus on conducting experimental studies to evaluate the effectiveness of artificial intelligence, educational robotics, and maker education in different settings. Creating guidelines on how to effectively introduce and use them in teaching and learning activities is also important. Moreover, emphasis should be placed on providing appropriate training to educators to effectively integrate them into their classrooms and on creating related educational resources. Future studies should also examine how students' background and characteristics influence the learning outcomes in such environments.

References

[1]   Feng X, Zhang Y, Tong L, Yu H. A bibliometric analysis of domestic and international research on maker education in the post-epidemic era. Library Hi Tech, 2024, 42, 33–53. https://doi.org/10.1108/lht-04-2022-0187

[2]   Dougherty D. The maker movement. Innovations: Technology, Governance, Globalization, 2012, 7, 11–14. https://doi.org/10.1162/inov_a_00135

[3]   Kwon B-R, Lee J. What makes a maker: The motivation for the maker movement in ICT. Information Technology for Development, 2017, 23, 318–335. https://doi.org/10.1080/02681102.2016.1238816

[4]   Dougherty D. The maker mindset: design, make, play. Routledge; 2013, pp. 7–11.

[5]   Halverson ER, Sheridan K. The maker movement in education. Harvard Educational Review, 2014, 84, 495–504. https://doi.org/10.17763/haer.84.4.34j1g68140382063

[6]   Shi Y, Cheng Q, Wei Y, Liang Y. Linking making and creating: The role of emotional and cognitive engagement in maker education. Sustainability, 2023, 15, 11018. https://doi.org/10.3390/su151411018

[7]   Hsu Y-C, Baldwin S, Ching Y-H. Learning through making and maker education. TechTrends, 2017, 61, 589–594. https://doi.org/10.1007/s11528-017-0172-6

[8]   Martin L. The promise of the maker movement for education. Journal of Pre-College Engineering Education Research (J-PEER), 2015, 5. https://doi.org/10.7771/2157-9288.1099

[9]   Sang W, Simpson A. The maker movement: A global movement for educational change. International Journal of Science and Mathematics Education, 2019, 17, 65–83. https://doi.org/10. 1007/s10763-019-09960-9

[10]   Kolb DA. Experiential learning: Experience as the source of learning and development. FT Press; 2014.

[11]   Chen L, Chen P, Lin Z. Artificial intelligence in education: A review. IEEE Access, 2020, 8, 75264–75278. https://doi.org/10.1109/access.2020.2988510

[12]   Mubin O, Stevens CJ, Shahid S, Mahmud AA, Dong J-J. A review of the applicability of robots in education. Technology for Education and Learning, 2013, 1. https://doi.org/10.2316/journal.209.2013. 1.209-0015

[13]   Krajcik JS, Blumenfeld PC. Project-based learning. The Cambridge handbook of the learning sciences; 2005, pp. 317–334. https://doi.org/10.1017/cbo9780511816833.020

[14]   Hmelo-Silver CE. Problem-based learning: What and how do students learn?. Educational Psychology Review, 2004, 16, 235–266. https://doi.org/10.1023/b:edpr.0000034022.16470.f3

[15]   Chen X, Zou D, Xie H, Cheng G, Liu C. Two decades of artificial intelligence in education. Educational Technology & Society, 2022, 25, 28–47.

[16]   Zhai X, Chu X, Chai CS, Jong MSY, Istenic A, Spector M, et al. A review of artificial intelligence (AI) in education from 2010 to 2020. Complexity, 2021, 1–18. https://doi.org/10.1155/2021/8812542

[17]   Talan T. Artificial intelligence in education: A bibliometric study. International Journal of Research in Education and Science, 2021, 7, 822–837.

[18]   Brynjolfsson E, McAfee A. Artificial intelligence, for real. Harvard Business Review, 2017, 1, 1–31.

[19]   Li D, Du Y. Artificial intelligence with uncertainty. CRC Press; 2017. https://doi.org/10.1201/ 9781315366951

[20]   Stone P, Brooks R, Brynjolfsson E, Calo R, Etzioni O, Hager G, et al. Artificial intelligence and life in 2030: The one hundred year study on artificial intelligence 2016. https://doi.org/10.48550/ARXIV. 2211.06318

[21]   Haenlein M, Kaplan A. A brief history of artificial intelligence: On the past, present, and future of artificial intelligence. California Management Review, 2019, 61, 5–14. https://doi.org/10.1177/ 0008125619864925

[22]   Duan Y, Edwards JS, Dwivedi YK. Artificial intelligence for decision making in the era of big data – Evolution, challenges and research agenda. International Journal of Information Management, 2019, 48, 63–71. https://doi.org/10.1016/j.ijinfomgt.2019.01.021

[23]   Chen X, Xie H, Zou D, Hwang G-J. Application and theory gaps during the rise of artificial intelligence in education. Computers and Education: Artificial Intelligence, 2020, 1, 100002. https://doi.org/10. 1016/j.caeai.2020.100002

[24]   Lin R, Zhang Q, Xi L, Chu J. Exploring the effectiveness and moderators of artificial intelligence in the classroom: A Meta-Analysis. Resilience and Future of Smart Learning, 2022, 61–66. https://doi.org/ 10.1007/978-981-19-5967-7_7

[25]   Lampropoulos G. Augmented reality and artificial intelligence in education: Toward immersive intelligent tutoring systems. Augmented Reality and Artificial Intelligence, 2023, 137–146. https://doi.org/10.1007/978-3-031-27166-3_8

[26]   Chang J, Lu X. The Study on Students' Participation in Personalized Learning under the Background of Artificial Intelligence. In: 2019 10th international conference on information technology in medicine and education (ITME). 2019. https://doi.org/10.1109/itme.2019.00131

[27]   Jin S-H, Im K, Yoo M, Roll I, Seo K. Supporting students' self-regulated learning in online learning using artificial intelligence applications. International Journal of Educational Technology in Higher Education, 2023, 20. https://doi.org/10.1186/s41239-023-00406-5

[28] Ouyang F, Zheng L, Jiao P. Artificial intelligence in online higher education: A systematic review of empirical research from 2011 to 2020. Education and Information Technologies, 2022, 27, 7893–7925. https://doi.org/10.1007/s10639-022-10925-9

[29] Ouyang F, Jiao P. Artificial intelligence in education: The three paradigms. Computers and Education: Artificial Intelligence, 2021, 2, 100020. https://doi.org/10.1016/j.caeai.2021.100020

[30] Chiu TKF, Chai C. Sustainable curriculum planning for artificial intelligence education: A Self-Determination theory perspective. Sustainability, 2020, 12, 5568. https://doi.org/10.3390/su12145568

[31] Scaradozzi D, Screpanti L, Cesaretti L. Towards a definition of educational robotics: A classification of tools, experiences and assessments. Smart Learning with Educational Robotics, 2019, 63–92. https://doi.org/10.1007/978-3-030-19913-5_3

[32] Angel-Fernandez JM, Vincze M. Towards a formal definition of educational robotics, 2018. https://doi.org/10.15203/3187-22-1-08

[33] Ospennikova E, Ershov M, Iljin I. Educational robotics as an innovative educational technology. Procedia – Social and Behavioral Sciences, 2015, 214, 18–26. https://doi.org/10.1016/j.sbspro.2015.11.588

[34] López-Belmonte J, Segura-Robles A, Moreno-Guerrero A-J, Parra-González M-E. Robotics in education: A scientific mapping of the literature in web of science. Electronics, 2021, 10, 291. https://doi.org/10.3390/electronics10030291

[35] Anwar S, Bascou NA, Menekse M, Kardgar A. A systematic review of studies on educational robotics. Journal of Pre-College Engineering Education Research (J-PEER), 2019, 9. https://doi.org/10.7771/2157-9288.1223

[36] Benitti FBV. Exploring the educational potential of robotics in schools: A systematic review. Computers & Education, 2012, 58, 978–988. https://doi.org/10.1016/j.compedu.2011.10.006

[37] Gunes H, Kucuk S. A systematic review of educational robotics studies for the period 2010–2021. Review of Education, 2022, 10. https://doi.org/10.1002/rev3.3381

[38] Papadakis S, Vaiopoulou J, Sifaki E, Stamovlasis D, Kalogiannakis M, Vassilakis K Factors that hinder in-service teachers from incorporating educational robotics into their daily or future teaching practice. Proceedings of the 13th international conference on computer supported education, 2021. https://doi.org/10.5220/0010413900550063.

[39] Schina D, Esteve-González V, Usart M. An overview of teacher training programs in educational robotics: Characteristics, best practices and recommendations. Education and Information Technologies, 2021, 26, 2831–2852. https://doi.org/10.1007/s10639-020-10377-z

[40] Pedersen BKMK, Larsen JC, Nielsen J. The effect of commercially available educational robotics: A systematic review. Robotics in Education, 2020, 14–27. https://doi.org/10.1007/978-3-030-26945-6_2

[41] Atmatzidou S, Demetriadis S. Advancing students' computational thinking skills through educational robotics: A study on age and gender relevant differences. Robotics and Autonomous Systems, 2016, 75, 661–670. https://doi.org/10.1016/j.robot.2015.10.008

[42] Chen X, Xie H, Hwang G-J. A multi-perspective study on artificial intelligence in education: Grants, conferences, journals, software tools, institutions, and researchers. Computers and Education: Artificial Intelligence 2020;1:100005. https://doi.org/10.1016/j.caeai.2020.100005.

[43] Chevalier M, Giang C, El-Hamamsy L, Bonnet E, Papaspyros V, Pellet J-P, et al. The role of feedback and guidance as intervention methods to foster computational thinking in educational robotics learning activities for primary school. Computers & Education, 2022, 180, 104431. https://doi.org/10.1016/j.compedu.2022.104431

[44] Eguchi A. Educational robotics as a learning tool for promoting rich environments for active learning (REALs). Handbook of Research on Educational Technology Integration and Active Learning, 2015, 19–47. https://doi.org/10.4018/978-1-4666-8363-1.ch002

[45] Daniela L, Lytras MD. Educational robotics for inclusive education. Technology, Knowledge and Learning, 2019, 24, 219–225. https://doi.org/10.1007/s10758-018-9397-5

[46] Socratous C, Ioannou A. A study of collaborative knowledge construction in STEM via educational robotics. International Society of the Learning Sciences, Inc. [ISLS].; 2018. https://doi.org/10.22318/cscl2018.496

[47] Screpanti L, Cesaretti L, Storti M, Scaradozzi D. Educational Robotics and Social Relationships in the Classroom. In: Makers at school, educational robotics and innovative learning environments. 2021. pp. 195–201. https://doi.org/10.1007/978-3-030-77040-2_26

[48] Eguchi A. Educational robotics for promoting 21st century skills. Journal of Automation, Mobile Robotics and Intelligent Systems, 2014, 5–11. https://doi.org/10.14313/jamris_1-2014/1

[49] Atman Uslu N, Yavuz GÖ, Koçak Usluel Y. A systematic review study on educational robotics and robots. Interactive Learning Environments, 2023, 31, 5874–5898. https://doi.org/10.1080/10494820.2021.2023890

[50] Evripidou S, Georgiou K, Doitsidis L, Amanatiadis AA, Zinonos Z, Chatzichristofis SA. Educational robotics: Platforms, competitions and expected learning outcomes. IEEE Access, 2020, 8, 219534–219562. https://doi.org/10.1109/access.2020.3042555

[51] Zhang Y, Zhu Y. Effects of educational robotics on the creativity and problem-solving skills of k-12 students: A meta-analysis. Educational Studies, 2022, 1–19. https://doi.org/10.1080/03055698.2022.2107873

[52] Chevalier M, Giang C, Piatti A, Mondada F. Fostering computational thinking through educational robotics: A model for creative computational problem solving. International Journal of STEM Education, 2020, 7. https://doi.org/10.1186/s40594-020-00238-z

Rina Zviel-Girshin and Nathan Rosenberg

Making to engineering: toward a personalized engineering robotics playground

Abstract: Using educational robotics playgrounds at different stages of development, children could evolve from the most basic making skills (MSs) to developing very sophisticated MS, which become part and parcel and one of the drivers of developing engineering skills (ESs). This process could, and should, start as soon as possible. The newborn instinctively starts to try/play/make, sharpening its MS, and in the process develops all its mental and physical strengths, including those that will become the beginnings of its ES. Creating a special engineering robotics playground can improve these processes both qualitatively and quantitatively while establishing the vital creative link between making and engineering (MS and ES) and internalizing best making and engineering practices.

Keywords: Education, educational robotics (ER), science, technology, engineering, and math (STEM), making, engineering skills, engineering robotic playground (ERP)

1 Introduction

Some argue that making and more planned engineering are as contradictory as induction and deduction. We beg to differ. Both are the two sides of one and the same coin, that is, lab practice or applicative engineering, which is at the heart of any, be it as novel as possible, engineering process. We hold as true that all engineering processes and each of their stages include both analytic (more theoretical and a priori) and synthetic (maker's tinkering and engineer's building the prototype in the lab) aspects. Moreover, they are indivisible. But the skill of using them in concert, dovetailing in the right order and ratio, is as rare as it is great. It could even be argued that this making skills to engineering skills (MS-ES) bridge is an indispensable one for a creative engineer.

What about the opposite direction? Are engineering skills (ESs) vital to a maker? Here the answer is more ambiguous. As making is a more basic skill, it is a precondition and a part of more complicated activities. However, the added elements in engineering are not necessarily a conditio-sine-qua-non for making. Let's clarify this with

---

Corresponding author: Rina Zviel-Girshin, Ruppin Academic Center, Central, Israel,
e-mail: rinazg@ruppin.ac.il
Nathan Rosenberg, Paralex Institute, Israel, e-mail: paralex.research@gmail.com

https://doi.org/10.1515/9783111352695-002

a simple analogy. The ability to read (or think, write, listen, etc.) is a necessary condition for gaining an academic degree. But having an academic degree is not a precondition to being able to read (think, write, listen, etc.). This analogy will help us further understand the sophistication of the relationship between making versus engineering. An academic degree is certainly not a prerequisite to the ability to read. On the other hand, education in general, and higher education in particular, will improve reading skills even in the technical sense and should dramatically improve the impact of reading and the benefits gained by reading when compared to a reader lacking education.

It is, of course, only a probabilistic argument for both directions. Education without being able to read is much less probable (practically impossible) than being able to read without higher education. Some of the more intelligent and successful people did not have the advantages of higher education. And yet, as Asimov (allegedly) has put it: "I am not a speed reader. I am a speed understander" [1]. So, the maker could also benefit from ESs. Yet we should remember, in this context, at least two caveats. First, the objective one, that some engineering practices could stifle the making process. Second, the subjective one, that one of the first principles and aims of the modern maker movement is to serve as an antithesis to the more formal approach like institutionalized engineering.

The chapter is organized as follows. We start by examining making and engineering separately and their interplay. Then, we discuss the skills and traits associated with making and engineering. Next, we describe the Engineering Robotic Playground (ERP) as infrastructure exceptionally suited for the development of MS-ES from the youngest to the oldest of ages. Following that, we describe various ERP programs, their participants, features, learning methodologies, and outcomes. We provide a brief overview on several of the ERP programs starting from pre-K to university. The chapter proceeds with a brief conclusion and discussion of future research directions.

2 Making, engineering, and manufacturing

Making, makers, maker-places, and maker-spaces could have at least two quite distinct semantics: movement and need-joy-activity. Making and engineering could be understood, each independently, in multiple ways. Depending on the meaning we give each of those terms their interaction could vary dramatically from no interaction at all to almost totally overlapping. Engineering could be viewed stricto sensu as only formal preplanned algorithmicized activity done professionally in a formal setting of appropriate organization and exclusively by holders of an engineering degree from an institution of higher education (and we can go on and demand that they will be members of a professional engineering association (guild) such as the American Society of Mechanical Engineers, American Society of Civil Engineers, or Institute of Electrical and Electronics Engineers. Then we can define making as all creative activity that does not meet those strict conditions, all or some. Had we taken those definitions as the basis for our analysis, it would have been obvious that the two are different and have nothing in common. And if we go further on that road and postulate that engineering is formal and organized versus making that is spontaneous and community-based, we could easily arrive at the conclusion that those two are beyond different – they are ideological enemies.

We hold a diametrically opposite view. Our feeling, based on many decades of experience in engineering and makers playgrounds, is that the two are so similar that they are almost indistinguishable. That approach, of course, is the result of a much wider and more realistic view of the meaning of the two terms. Without precluding others, that in other context and for other purposes would define the relation in another way (and sometimes would be right to do so), we here describe the making-engineering relation as making **and** engineering, and more precisely making **to** engineering (M-E), and even making **for** engineering and making **in** engineering. For the purposes of our research, and to maximize its practical conclusions and outcomes, we will hold that both making and engineering have a kernel of truth of a paradigm that is vital for engineering and beneficial for making.

Making and engineering have a lot in common, to the extent that it would be quite difficult to clearly distinguish between them. They are both creative and they are not theoretical. Their creativity is a practical one – aimed at making something that is immediately and directly useful. We would call those engaging in these fields utilitarians, but for their great enthusiasm that makes them much more than mere utilitarians. The common field of activities could be called productive creativity. The differences are small and vary, and yet, if we have to emphasize them, they could be described as follows: one is more formal, another more informal. One is more a priori research, another is more trial-and-error. One is more spontaneous, another more planned. One is more of a profession another more of a passion. One is more conformist, another more individualistic. The list can go on and on. But the fact is that we don't have to say which is which proves that really at some level we all understand quite well the differences between making and engineering.

Productive creativity is at its best when it incorporates an intricate ecosystem of both, to form an activity that has the advantages of both. This demands the right education and experience, preferably from the earliest of ages. It also requires a sophisticated organization and process to use both in an optimal way. But the huge advantages are worth the investment. Such a **Making-Engineering** ecosystem could be called MaGineering (or just magineering). So, the most creative and productive maker or engineer would be the magineer. Magineer is not to be confused with mage, magician or imagineer, yet the association is still very relevant. Interestingly, all those words come from the Proto-Indo-European root "magh" meaning "to be able, to have power," and "engineer" comes from the Proto-Indo-European root "gene" meaning "produce." So magineer is one having the better abilities to produce.

But if the essence (and aim) of these activities and their corresponding skills is to produce, aren't they just other words for manufacturing? They are certainly not manufacturing, although each in quite a different way. Manufacturing, defined as the transformation of materials and information into goods for the satisfaction of human needs is one of the primary wealth-generating activities [2]. But aren't both making and engineering, being a production processes, merely manufacturing? They are both transformations of materials and information into goods, for the satisfaction of human needs, but each has another aspect and another set of skills. Manufacturing needs skills for production in the strict sense. Making is production in a wider sense, having also the characteristics of individuality, enthusiasm, uniqueness, creativity, etc. Engineering, in the best sense, adds another level of progress to manufacturing, based on making. Engineering adds the layer of technological-scientific knowledge-oriented, critical and analytical thinking. It also adds the more technological-scientific creative imagineering.

Those are three distinct paradigms, although, ideally, they should be closely cooperating in one ecosystem. When this system integrates all three layers of making, engineering, and manufacturing, the term "magineering" could also include manufacturing (magineering = **ma**king-en**ginee**ring-manufactu**ring**). This kind of ecosystem needs a toolbox of skills for each of the layers and also some more integrative special skills for success in the more sophisticated multifaceted system (e.g., the ability to easily move from one layer to another, find the corresponding components, and use them in a timely and an efficient way).

2.1 Magineering process

The ecosystem of magineering is not rigid or linear. It is rather an evolutionary spiral of building a more and more intricate organization of abilities, traits, skills, active agents, and mechanisms reflecting the sophistication of the real world in which magineering operates. Still, we can delineate it to present a partial, noncomplete, yet very useful description of the model.

2.1.1 Making process

1. Experiencing the real-life needs and state of the art of products
2. Mentally playing with this knowledge
3. Imagining and brainstorming
4. Tinkering
5. Producing individual prototype
6. Feedback through dialog with the makers' community and general public

7. Joy and pride in the completed product (even when it is not perfect)
8. Sometimes advancing the product for wider production and use (optional)

2.1.2 Analytic-synthetic engineering creative process

1. Analytically – the engineer analyzes the existing knowledge (the problem, solutions, facts, processes, and procedures)
2. Synthetically – plays with it (constructing in his mind versions of mental mirror of it)
3. Analytically – transforms the numerous versions into a settled subset of inner structured knowledge
4. Synthetically – creatively imagining new knowledge based on what is learned and reconstructed
5. Analytically – brainstorms the problems with the new knowledge possibilities
6. Synthetically – constructs a new, better model of the problem in the context of the mental image of the problem domain
7. Synthetically creates a multitude of prototypes of solution
8. Synthetically – the new synthetic knowledge is evolving with more components becoming a working prototype to be tested
9. Analytically – he reviews his new knowledge hypothesis,
10. Analytically – tests in mental experiments in analytic mode.

The engineering process is an evolution of many stages of analysis and synthesis. First, creating the mental prototype (mental-making), then experimenting, receiving feedback, and remaking in a spiral of better prototype-synthesis at each iteration. The prototype evolves and changes while making, testing, and feedback. In addition to the prototype and product made, new engineering knowledge is created.

2.1.3 Manufacturing

1. Getting the prototype of the product
2. Researching and getting a sense of requirements, demand, market, and opportunity
3. Feasibility decision about upscaling
4. Concretizing the upscaling
5. Design of the product and its manufacturing
6. Building the manufacturing process
7. Feedback and improvement
8. Maintenance and growth

3 Making and engineering skills

The different skill sets should be analyzed both holistically (as a set or a gestalt) and on the basis of each skill separately. For instance, the skill of perseverance could be treated as one skill. Then we would talk to students about the importance of not giving up. We would also pose challenges that take a longer and longer time and support and encourage the student not to stop until the job is done. On the other hand, perseverance could be treated as an intricate organization of many (interdependent to various degrees) traits. It could include mental planning, the ability to see the whole picture, strong motivation, the ability to deal with failure, keeping high energy levels, the ability to ask for, and get help from others that will enable continuing the project for a longer time. In this more multifaceted approach for developing perseverance, we would work on many other skills, including such seemingly unrelated as best habits of eating and drinking, and organizing a better workplace, including details such as lighting, colors, music, etc. It could include skills and habits such as making lists (specifically prioritize tasks), concentrating on the more important tasks, working on parallel tasks, delegating work and trust others, calendars, timelines, and time planners, managing the physical and online working environment (specifically desktop), designing, and constantly updating the life cycle of the project, and many more.

It could be argued that in some cases those skills are linear and scalar, that is, they have a size. So, we can talk of 100% perseverance. That could be accurate when a project was successfully completed with no breaks. In other cases, it is a matter of degree. For instance, we can talk about the perseverance degree of a developer (very high, high, medium, low, and very low). Categories of students and engineers, defined by this trait, could be characterized as belonging to one or another category. In another approach to measuring a skill, and even defining it, it is more useful to see two more extreme traits, and the specific skill is a point on the scale somewhere between the two extremes. So, perseverance degree is a measure of where the engineer is on the scale between being a total very quick quitter and a blind-to-reality obstinate. Here the two extremes are by themselves a negative characteristic, too far in one direction. Other, less extreme, positive two poles of the scale could be adopted. In the case of perseverance those could be flexibility and being very sensitive to feedback, as one pole, and never-surrendering, as the other.

So, every skill could be defined by a pair of traits, each of them contrasted with the symmetrical opposite trait. So, every skill is actually the knack of finding the right equilibrium between two opposing poles (that can be either positive in themselves, like courage and caution, or negative like passivity and hyperactivity). This approach is a general one and is applicable to every engineering aspect. This is an approach falling under the auspices of the golden mean method, usually attributed to ancient sages like Aristotle, Buddha, or Confucius [3], although, actually, the approach is much wider and more general, present in many teachings and cultures, beyond the teachings of one man.

In magineering, paraphrasing Aristotle's works, engineering traits could resemble an ontology of categories, where for each there are the two extremes of too-little and too-much, and the golden mean of good engineering between them. We present some examples in Tab. 1.

Tab. 1: Engineering traits divided into major categories by their strength.

Engineering strength	Deficiency	Mean	Excess
Team relations	Introvert	Ambivert	Extrovert
	Disloyal	Loyal	Too loyal
	Not participating	Contributing	Overpowering
	Silent	Communicative	Loquacious
Confidence	Timid	Brave	Reckless
	No initiative	Creative	Mad ideas
	Subservient	Independent	Tyrannical
Planning	Improvisor	Knowledgeable	Pedantic overplanning
	Short-sighted	Realist	Dreamer
Energy	Hypoactive	Energetic and active	Hyperactive

Approach using different constructs could be that of engineering strengths, where each strength should be used to the right degree so that between strength's overuse and underuse, the engineer has to strive for optimal use of each of his strengths [4].

The classical examples of the skills of optimal balancing, given all circumstances (like aims or resources), would be
- creativity as a balance between overreaching, too creative brainstorming, and conservative, noncreative thinking
- good goals are a balance between overpromising overoptimistic hopes and too little, less-than-needed afraid-to-try activity
- good team hierarchy is a balance between chaotic anarchy and dictatorship
- good teamwork is a balance between extreme task orientation and people orientation
- good theory-practice balances between just theorizing and just trial without any forethought

Also, every skill is not a frozen static state but a process of learning and improving this aspect. The skills are overlapping, interconnected, and have multiple meanings in different circumstances. The names and ontologies could be different (based on different axioms, models, viewpoints, or just incoordination) and confusing, a fact significantly hindering the attempt to systematize this basis for creative productive activity.

4 Engineering robotic playground

ERP is exceptionally suited as the infrastructure for MS-ES development from the earliest to the oldest of ages. The educational robotics project is a bridge between MS and ES. It is so enjoyable and challenging that the motivation to learn MS-ES is at its highest. We have created (in various programs) such ERPs for all the stages of this evolution – from pre-K through primary and then high school in the context of K-12 education and even in our engineering school for undergraduate students.

Making is a natural activity. Children enjoy making and tinkering. They like to think (and perform mental experiments) and then build and implement their thoughts. Children are curious and like to try new things. However, during the years they spend in educational settings, it seems that they lose this interest in making. Traditional educational settings and institutions encourage a priori thinking, logic-mathematical-algorithmic activities, where participants are encouraged to perform well-defined steps. However, engineers in real-life play, tinker, and build not less than they think and plan. Moreover, the best engineering, the creative engineering, is much more about experimenting with building and making than realizing a predefined design. This culture of imagining, invention, and thinking is an important engineering activity and can be learned and perfected through project-based learning via educational robotics. The lab and playground are the places for real engineers and even more so for the future ones. The engineer has to have the vital lab skill of research and development (R&D), the process of redesigning mental picture, prototyping, playing with it, tinkering with it, building another variation (that will more and more stray from the original, as in the making the design is improved), getting feedback, learning from failure, and starting the next, better turn of the engineering spiral.

The idea of the MS-ES playground is not new, and the first such playground was established some 20 years before the word "robot" was first invented by Karel Capek, certainly before any educational robot was available. It was the Maria Montessori's Casa dei Bambini [5]. Maria Montessori created her Casa dei Bambini in Rome in 1907 as part of the Franchetti Foundation, known later as Centro Studi e Formazione Villa Montesca, and today as Fondazione Centro Studi Villa Montesca. Her idea was to replace the traditional, crashing-student-curiosity-and-instinct-of-making school by a project-based maker studio with the aim of developing MS-ES. After her pupils, many of whom were lagging in their development and education before starting the Montessori school, and some were even diagnosed as retarded, won mathematics Olympiad, the Montessori model became extremely popular. According to Montessori [6], "Free the child's potential and you will transform him into the world."

Our MS-ES approach integrates the ideas of Montessori, Dewey, Vygotsky, and Pappert to create a creative playground that not only teaches engineering but also develops skills, traits, motivation, and engineering personality, all wrapped in joyful playing, overcoming failures, and conflicts, progressing from making out of curiosity to R&D of the highest level. The Montessori model is that of maximal freedom and curiosity-driven, project-based tinkering and making, but channeled into intensive learning through R&D of engineering projects, as close to real life as possible for the individual student.

Seymour Pappert's work is extremely important in our field [7, 8]. It could be argued that he not only laid out the theory and principles of ERP as early as the 1960s, but actually built the first ERP. His lab at the Massachusetts Institute of Technology (in cooperation with Marvin Minski) was always at the spearhead of robotics in education and making as a vehicle for science, technology, engineering, and math (STEM) skills. Though he concentrated on mathematical skills, his idea of empowering children to experiment, explore, and express themselves through building and programming robots, and his great theoretical and practical work certainly entitle him to be named "the father of ERP."

A crucial factor in ERP is Vygotsky's zone of proximal development (ZPD). According to this visionary of the sociocultural environment for constructivist individual growth, the learning environment should be customized to the individual student's ability (ZPD) [9]. Notwithstanding the tragic circumstances of his life in postrevolutionary USSR, in poverty and destitution, he was very optimistic and believed that with the right educational environment ("scaffolding") the ZPD could be stretched, and the student could evolve much faster than could have been expected. The scaffolding adds to the variety of the student's mental tools, thus widening the ZPD.

Following John Dewey, the environment is one of continuous constructivist dialog. Following Seymour Pappert, who as far back as the 1960s created educational robot (the turtle) and the educational LOGO programming language (with Wally Feurzeig and Cynthia Solomon), the robotics is the basis for the making-engineering play (being actually R&D project). The robotic makers-engineering lab is the quintessence of all those educational approaches that, though they are at the center of the pedagogical consensus, have not actually conquer the K-Uni educational institutions. We believe that this lacuna is not due to extraneous reasons but arises from the lack of a practical alternative to more traditional teaching. The ERP was created as such an alternative.

The magineering education should start from the earliest age possible, by immersion in ERPs and doing R&D of robotics engineering projects appropriate for the age, for the context, and for the individual student. It would be optimal for it to seamlessly continue to other levels all through education, including higher education, and preferably even afterward, as a very important tool for further professional development (In cooperation with the employer or beyond it). The MS-ES approach has allowed us, for the last 12 years, to create a variety of ERPs realized in various programs for different age groups. Hereunder we shall bring some examples of those ERPs.

5 Engineering robotic playground organic process

ERP is the best platform for building the bridge from MS to ES, as it features a set of characteristics, elements, and mechanisms that could be called "organic" and are especially conducive and suitable for this goal:

- Organic nonalgorithmic environment in process (how it works, projects' R&D) and result (outcome)
- Nonblind trial-and-error making
- Model (maximal knowledge) of the problem and the world and experimentation on gestalt (paradigm)-driven prototypes as an engineer and then as a maker
- Agile immediate prototype widest usage of making
- Fuzzy human language constructive creativity through human dialog (even when automated) as part of the lab and maker studio teamwork
- Analog versus digital widest and deepest knowledge and universe of hypotheses while building and trying out the prototypes
- Play-work-create-use-socialize merging in the lab and on the playground
- Checks and balances, dialectical competing Darwinist evolution of the making process and engineering project
- Competition among makers' teams and other stakeholders
- Intricate system of rewards
- Multilevel projection and bubbling of ideas and implementation through the making process
- Dialog is a way to formulate explicit specific clearer description which is also a constructive prescription and so a posteriori analysis of the making process
- Engineer between (amalgamation of both) scientist and maker

The resulting organic system has numerous advantages such as acceptability, evolution, interoperability, maintainability, rapidity, reliability, reuse, robustness, supportability, testability, understandability, user experience, and validity.

What makes the playground, project, or system organic? We will first try to answer the question in very simple (even simplistic) terms. In algorithmic programming, the programmer must find a solution to the problem before he starts to code. Once he has the complete noncomputerized solution, he creates an algorithm to implement his solution in pseudocode and then in code. This solution produces the unique, exact, correct answer, and it does so always. In organic approach we don't have *the* solution or even *a* solution, not even after the coding is complete. We just create a copy of the world or mechanisms for operating in the world, and let the program work, interact with the environment, learn, and thus become useful, not necessarily giving the most correct, or even nearly correct solution. The advance from more incorrect to less incorrect is not only acceptable, but most of the time the only possible strategy.

If we must characterize the organic method, in context of this book, in one word, it would be evolution. In two words, it would be Darwinian evolution. The evolution

Organic SE Lifecycle

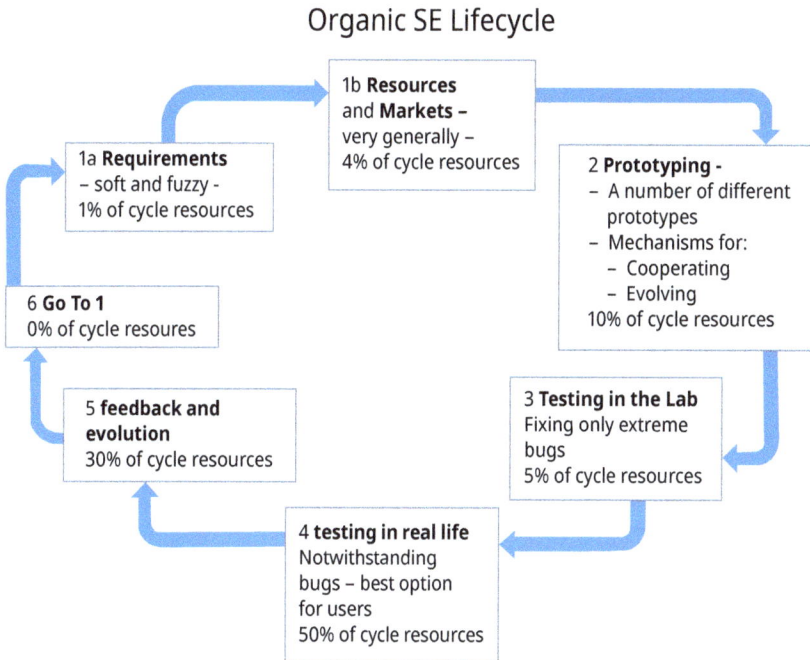

Fig. 1: Organic ecosystem lifecycle.

from simplest system to more advanced, organic ecosystem could take a very long time, and even in the initial development stages, it could include many iterations of prototype improvements [10] (Fig. 1).

6 Engineering robotic playgrounds programs

The ERP programs, though the same at their core (the paradigm and method level), differ greatly as they are constantly customized to students. Beyond individual customization, high-resolution micro-adjustment fine-tuning, many programs were created, each tailored to a different group. The programs differ in all aspects, as for every target group all mechanisms are reinspected and adjusted. That includes the physical environment and the tools chosen. Among the great variety of robotic hardware/firmware/software systems used as educational platforms, the following ones are used: basic electrical, electronic and robotic kits, different LittleBits kits, Meccano, several bots of Makeblock, Lego Education products, like WeDo, MindStorms (NXT and EV3) and the latest version Spike, Pololu, Arduino, Raspberry Pie, and other proprietary systems.

The participants vary by age, education level, context, or other characteristics of the participants. Programs were created for different age and education levels
– pre-kindergarten (pre-K)
– kindergarten
– grades 0–1
– grades 2–4
– grades 5–6
– middle school
– junior high school
– senior high school
– undergraduate engineering students

Other ERP programs were specifically designed for participants belonging to groups such as
– ERP workshops and hackathons
– special needs students
– immigrants
– woman
– minorities

We will hereunder provide a brief overview of some of the programs.

6.1 ERP workshop and hackathon

Recently, a very intensive makers' fest-atmosphere hackathon with engineering workshop and playground environment has proved to be extremely popular. We adapt them to the specific group of participants in every aspect: goals, content, level, challenges, and spirit. One example would be the science day summer activity (happening once a week approximately and involving many hundreds of participants) is a program in which participants participated in 1-day ERP hackathon where they are exposed to a variety of different robots, from industrial arm robots to Arduino-based robots, LEGO programmable robots, and LittleBits makers kits (Fig. 2).

During this hackathon, students were divided into teams and each team chose a making project, which is also an engineering real-life project. For example, building a robotic conveyor process that takes part from the factory warehouse to factory floor using a system of robots.

Fig. 2: Participants of science day summer activity using Industrial ScorBot ER-4U arm robot.

6.2 Pre-kindergarten and kindergarten ERP

6.2.1 Pre-kindergarten

The early childhood in our context is defined individually, developmentally, and functionally. The criteria are first and foremost the ability of the toddler to play. Then the attitude of their caregivers (primarily parents), and the feasibility and practicability of creating an active home-building playground, which is the age-appropriate version of a maker's and engineering playground. This stage lasts at least until the child can go to their first out-of-home educational setting. Of course, the parents can (and should) continue with this home-building playground in parallel with other educational settings their child participates in. The home playground is evolving, but it can continue to serve as great support and development of MESs for many years as well as serve other very important aims such as parent-child relations.

At this stage we are not involving building robots. Although it is usually a closely supervised activity by caregivers, safety considerations, as well as developmental factors, call for different building tools and materials. We chose, for instance, specially adapted LEGO (with building blocks much larger than usual) which offers a plethora of building alternatives and thus involves all the maker's skills. It could also evolve into the toddler being involved in the parent's building of simple science toy sets chosen by us like very rudimentary electric cars, windmills, and dinosaurs. And the preschooler is taking a more and more substantial part in this maker's process. The parent explains slowly and demonstrates all the steps, many times assembling and disassembling the toy, encouraging the child to take a more and more active part in the build.

6.2.2 Kindergarten

Early childhood program (pre-K and K) running since 2016 has involved over 2,500 children in several kindergartens. The program uses project-based learning, motivates children to be creative, and promotes inquiry-based maker's project. This program, approved by the Ministry of Education, is a compulsory, free-of-charge program for kindergartens in formal education. In kindergartens a special robotics lesson was added once a week. A mediated learning approach that included both direct instruction (short lectures and multimedia demonstrations) and open-ended, student-directed inquiry (working in small groups to solve problems) was employed.

At the end of the school year, all teams received a final project that they were supposed to build using robots. This final project was on specific topics such as how robots can help humans on the Moon, how robots can be used to help domestic and wild animals, or how robotic devices can be used in a child's room. Afterward, all participants were invited to a Robotics Day, where they presented their work to other children, teachers, family members, and local authorities [11].

6.3 School ERP

6.3.1 Primary school

Primary school ERP is starting with short and simple but already robotic R&D projects. They include individual as well as teamwork. The projects include making both the hardware and software – both building the robot and programming it. With time the projects become larger, the challenges harder, teams bigger, and explanations by the instructors longer and deeper. In the beginning the emphasis is almost exclusively on building motivation and excitement through the joy of playing and making. Mistakes are less pointed out, and in general, not much importance is placed on the result. The children understand that their efforts as makers are by themselves the great achievement to be treasured. Their environment, especially their parents, is encouraged to praise their children and even exaggerate their enthusiasm about their projects. Although it is a continuous process of educational progress, in general, our programs consider three levels of primary school ERPs: grades 0–1, grades 2–4, and grades 5–6.

In one of the programs in grades 0–1, in the elementary schools robotics lessons were added to the first-grade curriculum. Initially 12 first-grade classes in three schools from diverse socioeconomic and ethnic backgrounds were involved. After several years the program was extended to 20 first-grade classes in six different schools. At school each lesson lasted two academic hours per week for the duration of the school year. The main equipment was the LEGO® Education WeDo 2.0 kit specially designed for elementary schools.

Open-ended, student-directed inquiry consisted of students working in pairs to solve problems posed as programming and building challenges. Some of the challenges were proposed to them by the instructor and were well-defined, and some were provided by the instructor but were loosely defined (intentionally leaving room for variation in problem solving) and some were their own creations (Fig. 3).

Fig. 3: A presentation of a loosely defined dog project.

In one program, during the lesson, the class was divided into two groups: one stayed in a regular classroom with the regular teacher and the other went to a science classroom, where a science teacher, who had completed training in the field of robotics, helped the children perform a robotic activity or solve a problem in the field of robotics. Each half of the class was later divided into teams to work together on assignments. Each half of the class employed a mediated learning approach that included both direct instruction and open-ended, student-directed inquiry. Direct instruction included short lectures and/or multimedia demonstration of a concept, principle, or model and projects, problems, or activity assignments. Open-ended, student-directed inquiry consisted of students working in pairs to solve problems posed as programming and design challenges (Fig. 4). Some of the challenges were provided to them by the instructor and were well-defined, some were provided by the instructor but were loosely defined (intentionally leaving room for variation in problem solving), and some were their own creations.

Fig. 4: Robotics play area at school in science class (K'Nex equipment).

In elementary schools after the lesson was over, the robotic kits were stored and were not available to the children until the next robotics lesson the following week. There were no additional robotics activities during the school week [12].

At the end of the school year, each team had built the final prototype. They presented the project's final implementation in hardware (robot) and the code controlling the final robot. These final project outcomes included the hardware, the software, the model of the system working in context (simulating real life environment), and the poster explaining the details of the project. All were presented by the team and the instructor at the exhibition of the projects at each of the schools, before a wide forum. Later, all schools that participated in the project were invited to the "Robotics Workshop" activity at Heffer Valley Culture and Science Center. They presented their projects and were exposed to a much greater number of other children, teachers, parents, and local authorities.

6.3.2 Middle school

Middle school program, which has been running for many years and has included many thousands of students, takes place in a special Science Center with extremely well-established labs equipped with a variety of robots and dozens of instructors and mentors (Fig. 5). Throughout the school year, participants join age- or topic-appropriate groups once a week for various activities. These activities involve a wide range of robots including LEGO WEDO, LEGO Mindstorm, Arduino, and robotic drones.

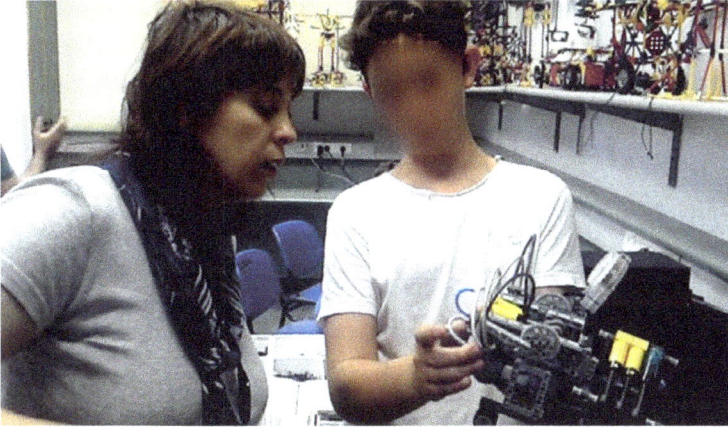

Fig. 5: Middle school participant presenting his LEGO artifact built by him.

6.3.3 High school

High school programs are aimed at developing and improving MS ES skills of high school students. One such program that is exceptionally interesting for this age group is intended for teenagers at risk in boarding schools. They were lagging in their STEM achievements but after attending 10–12 workshops of MS ES ERP they had a great leap in motivation and STEM achievements. Most important, most participants claim that they now intend to study hard with the aim of becoming robotics engineers. Their objective progress was closely monitored by the instructors, who found a very significant improvement in ES MS skills.

6.4 Undergraduate engineering students

6.4.1 Junior college model school

Junior college has existed since 2012 as a college-preparation school for gifted children. In this college, small groups of gifted children are taught STEM and English, adhering to the national curriculum published by the Israeli Ministry of Education, though not necessarily in the chronological order and age of the standard program.

It is based on an organic non-algorithmic paradigm for education [10]. A flexible model of the organic paradigm allows us to develop real-life information and communication technologies (ICT) learning environments which are much smarter and to a much higher degree simulate the most vital aspect of the teacher's expertise – the pedagogical know-how.

In 2018, the Rubin family donated money to the lab to create a new "Robotic Junior College Program named after Nehemia Rubin." The purpose of this program is to open the field of robotics education to the youth at risk and new immigrants to increase the academic achievement of pupil in STEM and to improve social mobility in the future.

6.4.2 Montessori-Dewey laboratory experimental school

MODEL (Montessori-Dewey Laboratory) School is a pilot school under the auspices of the Ruppin Academic Center. It was established in 2012 with a very small number of students and programs and has grown substantially since then. MODEL School is a unique attempt to create a synthesis of the two major paradigm-shifting educational schools, adjusting them to the realities of the twenty-first century, with a stronger emphasis on creativity and innovation, AI and robotics, engineering, twenty-first century skills, netizenship, and advocacy of children's rights and agency [13]. It is much more intricate in goals and structure and more ambitious in the aims. It incorporates the tremendous body of knowledge and know-how form of research and practical results of more than a century of education (dyohypostatic science and practice).

The school consists not only of classes but also mainly of labs (Fig. 6). One lab is the central hub, and then there exist a cluster of STEM labs, an infrastructure of same-locality educational institutions for all levels and ages of students pre-K-Uni: pre-K-K robotics lab in kindergartens, primary schools robotics lab, 4–12 robotics lab in the Science Center, AI and robotics lab in college. The younger students study, research, and develop college-level scientific endeavor projects in the college-level engineering lab. Because of their age, they enjoy a much freer atmosphere, more verbal communication from their instructors and much more monitoring and mental support.

Fig. 6: Junior college participants exploring different types of STEM projects.

In the spirit of Dewey, engineering and scientific dialog is at the center of the student's life much more than his work on experiments and projects. Theory is very important, even if it is an engineering theory. As a young engineer the student is constantly immersed in the process (an endless cycle of prototyping: requirements-design-implementation-testing-feedback-improvement). It is the triality of the three tech-creative-innovation hypostases: theory-engineering-technology (know-how). This is a more sophisticated version of Dewey's practice-theory-practice arc.

The headquarters lab is situated inside a college engineering school and serves students of all ages. Its open space is built as a number of connected spaces: theory and meetings (tables and chairs with whiteboards, projectors, and screens for presentations and lectures), makers' space of robots building and programming, robodrome, entertainment, rest, library, and outside activities with wide loans and inner yards.

The didactic and pedagogical philosophy at the heart of the school is an organic approach. It is a constructive synthesis of the two seminal educational paradigms – the revolutionary approaches of John Dewey and Maria Montessori. Though both are very similar in their anti-frontal-teaching basic approach, they are sometimes seen as contradictory. Both advocate the crucial role of the student, emphasizing less teaching and more learning. Like Vygotsky, they are both constructivists, that is, knowledge and character are not taught but individually constructed by the student in their own mind, in their own way, and at their own pace. Yet, one is more about individual skills, usually more STEM, and the other is more about moral and social skills, focusing on the development of a responsible and moral adult member of society. Like Vygotsky, one is more group-oriented, the other more individual. Still, we don't see an oxymoron here but rather two sides of the same coin. With an organic approach, the seeds of truth of both could, and should, be incorporated into the school.

All those founding fathers of constructivist modern education created a novel model of Laboratory School and realized it in a real-life pilot school. John Dewey created his Laboratory School at the University of Chicago in 1896 [14]. This evolved into today's highly successful University of Chicago Laboratory Schools, with more than 2,000 students, with an emphasis on continuous education starting with pre-K and K-12 that painlessly evolves into higher education. High school students take university courses as part of their studies. The University of Chicago Laboratory Schools are rated fourth in the United States.

Maria Montessori created her Casa dei Bambini in Rome in 1907 as part of the Franchetti Foundation, known later as Centro Studie Formazione Villa Montesca, and today as Fondazione Centro Studi Villa Montesca. After her pupils, many of whom before starting the Montessori school were lagging in their development and education, and some were being diagnosed as retarded, won the mathematics Olympiad, the Montessori model became extremely popular. The *Journal of Montessori Research Global Diffusion of Montessori Schools* report states that it found at least 15,763 Montessori schools in 154 countries worldwide, with the largest number of Montessori schools in the United States, China, Thailand, Germany, Canada, and Tanzania; and the United States, Thailand, the Netherlands, and India have the largest number of government-funded or public Montessori programs.

One formulation of the Montessori principles on which this ERP is based could be:
- recognizing the interests and needs of students, allowing their personal choice;
- reinforcing self-motivated, active, and autonomous, as well as collaborative, learning [15];
- individual, student-led pace and scenario, including learning by mixing students of different ages;
- facilitating the manipulation and understanding of materials;
- developing self-control;
- respecting freedom and teaching responsibility;
- encouraging creativity;
- encouraging trial-and-error techniques in tasks, among other characteristics.

Vygotsky created the Experimental Unified Work School in Moscow in the 1920s with the aim of applying his pedagogical approach of encouraging creativity and providing mental support for individualized constructivist learning. The school served as a model for his theories and as a pedagogical lab for testing and refining them.

6.4.3 ERP robotics undergraduate course

Since 2019 each year around 200 students who study electrical engineering and computer engineering have taken part in this course.

A cornerstone project course in the junior year of the Electrical and Computer Engineering Department, named "Engineering Skills and Thinking," is a stepping stone and simulator toward the final capstone project that students have to implement during their last year at the faculty of engineering. The course learning objectives are to provide the students with familiarity with the engineer's role as a problem solver and innovator and develop systemic reasoning alongside engineering and scientific thinking, and result in professional relatability with the role of an engineer. Prior to the course commencement, an educational team of eight instructors from different fields and backgrounds, including computer science, software engineering, computer engineering, and electrical engineering defined a list of skills essential for future electrical and computer engineers.

The course is 1-year (two semesters) long. In the first semester, the course focuses on fundamental electrical engineering concepts, basic programming (using Arduino IDE) and utilizing actuators and sensors. In its second part, during the second semester of the academic year, the students are required to plan, design, and build a "line follower" robot that is able, while tracking a narrow black stripe, to identify blocks on its way and report their color. In that part of the course, students learn about different phases of hardware and software-based design projects, while gaining experience with writing technical documents, abstracts, and concluding reports, as well as presenting the results to different audiences at various stages of the project. They acquire experience in mentored and independent team work. The students understand principles of the engineering design process and initial concepts in the management of engineering projects. Evaluation criteria are based on the final presentation and team competition results, awarding the best designed robot a special trophy. Out of the extensive toolbox of essential skills of the twenty-first century engineer, the course focuses on some of the more fundamental ones.

Students are gaining crucial engineering knowledge, habits, know-how, and traits:
– reinforcing engineering thinking and systemic understanding
    – students are required to design, build, and test a complete system
– acquisition of soft skills
    – oral and written communication
    – effective subject presentation
    – teamwork experience and self-study
– motivation enhancement

During the first semester, students participated in weekly lectures and mandatory lab activities. In-between lectures and laboratory sessions, students had to write and submit a preparatory report, which required self-studies and, in some cases, a small investigative project. Each lab project assignment had a list of tasks and problems as standalone tasks, which can be called mini-projects. The list of tasks was deliberately designed in a way that resulted in the majority of students not completing it entirely during the 2-h lab, requiring them to complete the lab assignments at home (the students were allowed to take the required equipment and robots with them). All implementation recordings and tests results had to be submitted to the course site. A wide variety of output forms were required and allowed: from the designed circuit (schematics or "Fritzing" sketches) and code to recorded short movies demonstrating the working solution. In the 10th week of the semester, students had a preparatory effective teamwork essentials workshop. Following this workshop, students were divided into teams of four members. Each team consisted of a team leader, a system engineer, a software engineer, and a hardware engineer. Teams started to work on multifaceted assignment that required building and testing a complete robot, writing the solution in code and documents, presenting the final product, and participating in a final competition.

The final competition is planned for the end of the second semester. Ten teams of four students are developing a prototype of the robot toward the final competition. They received a list of requirements and a kit of sensors they are allowed to use. Since we wish to encourage creativity and have a diversity of projects, each team can make up to two replacements within the basic kit without "paying" for it. Additional improvements are allowed with some penalty (points from the grade). Throughout the process, the teamwork is continually supported by the senior mentor, providing the guidance required for each team member to know their role, take responsibility for that role, and analyze the group processes in teamwork.

The preliminary results show high student involvement and satisfaction from the first half of the course. According to surveys conducted during the course students loved the idea of meeting different faculty members as course instructors during their first year. They found the course to be interesting, exciting, and demanding, improving their technical and soft skills. Student feedback was clearly positive regarding the project-based learning approach and accompanying "hands-on" activities.

The targeted learning outcomes are for the students to understand and experience basic principles of developing a complete system, understand what critical thinking is and how to approach every task and problem with an un-biased, analytical approach, gain practical knowledge of the engineer's role as a problem solver and innovator, experience and understand how to work as a team, realize and understand that training and self-learning never end, know how to write project documents and presentations, and be able to effectively present the project orally to different audiences from teachers to experts and the community.

6.4.4 Undergraduate robotics lab

The Knowledge Engineering and Robotics Lab provides the infrastructure for a multidisciplinary group of researchers and many students interested in specializing in the area of robotics. The Organic Knowledge (OK) approach to robotics sees robot as different from other computerized systems in that the robot is smarter, more involved, and more human-like. The OKbot has human expertise, sensors, and actuators, is more interactive with the environment and especially with humans, and has mechanisms for proactive learning and evolving.

Additional areas of the research are:
– STEM education
– educational software
– robotics education
– smart systems
– crowdsourcing and collective intelligence
– programming languages and environments

The *Gestalt-Multiplex-Layering* combination ERP lab project includes gestalt – a deeper model of the expert knowledge and reasoning process; multiplicity – simultaneous use and cooperation of different and conflicting approaches; layering – use of a hierarchy of independent layers of control and processing, through which the input and intermediate results are propagated. The independence of each layer enables the implementation of different approaches at different layers. The hierarchical layering of control and abstraction of lower by upper layers enables the cooperation and solution of contradictions arising from the use of a variety of different approaches. In very broad, plain terms, at each layer, there is a small knowledge system controlling, generalizing, and inducing the cooperation of different approaches in a larger knowledge system of the next layer.

The students' projects are at the highest level of engineering R&D and result in making a multitude of working prototypes. Some are described below:

6.4.4.1 REbot – recycled robot

REbot v1.01 is one of the robots built from recycled toys and industry leftovers in recycling ERP (Fig. 7). Building sustainable robots from recycled parts sounds like an extremely modern concept. Actually, it is not. The idea of maker kit of this kind was at the heart of the ideology of tinkering and making for a very long time. In 1898, in Liverpool, England, Frank Hornby started his very successful company producing construction kits from scrap. It was called Mechanics Made Easy since 1901 and Meccano since 1907. Similar kits were produced since 1913 by Alfred Carlton Gilbert under the name Erector. Such kits became the standard for makers and young engineers (actually their fathers too) all over the world. As far as behind the iron curtain, in communist USSR, they were in almost every household under the name Constructor. Most evolved with time into robotic construction kits.

Fig. 7: One of the robots built from recycled toys and industry leftovers in recycling ERP.

6.4.4.2 Dr Mec – fun educational assistant

The Meccanoid-based Dr Mec was built as a study-gamifying tool. It is based on the Meccanoid G15KS Personal Robot that comes with over 1,000 phrases and voice recognition. The robot is anthropomorphic making the young engineer both feel like he created a friend and cooperate with it as a peer (Fig. 8). The student can customize it, create his own dialog, and make the robot move, dance, and talk. It was used, for instance, during physics lessons in physics lab.

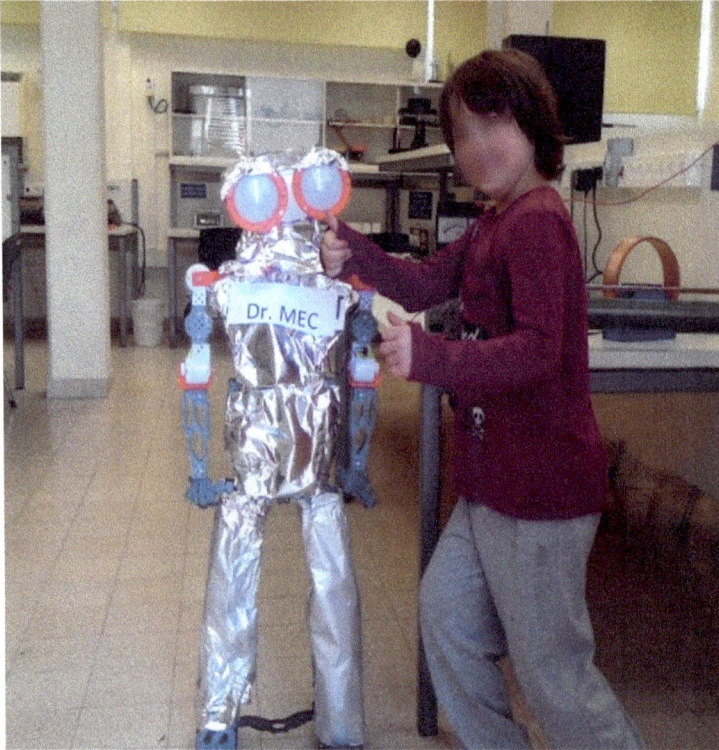

Fig. 8: Dr Mec ERP project.

Educational advantages
– Human-like input (voice and motion recognition and recording
– Learning by example
– Human-like output (human voice, dance, exercise, telling jokes)

6.4.4.3 MINbot

Marvin Minsky's Society of Mind is realized in a self-organizing society of small robots creating a robotic ecosystem. Among their abilities are

– communication
– cooperation
– mapping and positioning

6.4.4.4 MARINAbot marine robot

The marine robot, Marinabot, is a robot that has many strengths (Fig. 9). It is specifically engineered for use in water environments such as seas and oceans. It has several features:

– OK-based
– autonomous
– intelligent
– proactive
– sociable

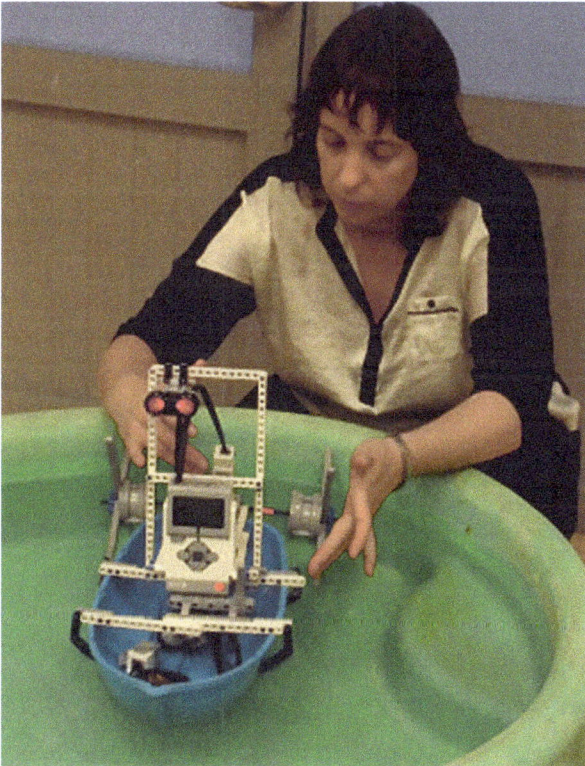

Fig. 9: MARINAbot testing.

6.4.5 ERP for capstone engineering project

In the last year of engineering school students create teams to engage in a capstone engineering project. The level of difficulty of this project is very high, as the aim is to engage the students in R&D that closely resembles the real-life engineering work they will do after graduation. Each team is closely accompanied and monitored by a multi-disciplinary team of faculty and industry professionals. The outcome of the project should be a working prototype of a novel advanced product created through a significant investment of imagination, innovation, time, and effort matching the best creative makers' and engineering projects. We present here some of the projects.

6.4.5.1 Assistobot

General assistant accompanies the user wherever he goes. It is a universal platform for a wide variety of uses, from gathering information to teaching, from telecommunication to telecontrol. It has many sensors and actuators built-in, but it is also modular, enabling customization by adding more sensors, actuators, and communication channels (Fig. 10). It can identify the user and open one of its drawers that contain something for this user only. Thus, it could become very useful in such tasks as medication dispensing:

– Arduino and RasberryPi-based Smart Assistant Robot
– autonomous
– self-recharging
– intelligent
– social assistant

6.4.5.2 Nannybot

Its function is to accompany the toddler and young child. The robot closely follows them but can also tempt the child to follow the robot. It has many features to attract the child's attention like flashing colorful lights, sounds, and an antenna to pull. It looks like a giant ladybug. One use is for exercise, to make the child move and run after the robot.

6.4.5.3 Agrobot

This is a robot used for agricultural use. It moves in the field or orchard and can gather data from numerous sensors like humidity, temperature, light, wind, soil quality, and visual images through a camera. It streams the data into a remote database to be used in an agricultural expert system.

Fig. 10: One of the models of Assistobot.

6.4.6 OOP ERP for engineering students

This LEGO-based ERP for undergraduate engineering students is an object-oriented programming (OOP) lab workshop. This project was specially designed to assist engineering students studying OOP languages and technology in better understanding OOP principles, abstraction, and building a complete system.

Among the aims of this ERP were

– demonstrating the difficulty of moving from the phase of specification and design to the phase of system implementation
– understanding the differences in ways of thinking according to the level of abstraction and detail that a task requires

Each academic year, several groups of students (35–50 students in each) studying OOP based on the C++ programming language participate in this ERP. In this ERP a spe-

Fig. 11: Participants of LEGO-based workshop for OOP who discuss their solution and "make" a prototype of the task.

cially designed organic approach to education is used. This approach integrates "making" principles and active learning in the OOP course. This approach could be seen as an organic version of problem-based M-E learning.

During the LEGO activity, students were divided into groups of four students (Fig. 11). Each group drew a project out of a lottery of tasks (such as building a house from LEGO bricks). The students were asked to design a relatively complex system when they were given a definition of the most basic unit (a class and an instance of the class) – the "Lego block." Students assembled a prototype using LEGO kits learning about OOP principles such as entity-relation, encapsulation. and inheritance. Later, they had to provide a product description using the "Lego Block" class and new classes derived from it or used as inner objects (including it). They were asked to present their design and explain the basic classes they would have built to solve the task. Each team was required to submit an OOP characterization/specification document of their solution.

6.5 Young woman engineers ERP

There is a broad consensus regarding the problem of female engineers – both concerning the depth of the crisis and the urgency to solve it.

Women could make a huge quantitative, and above all qualitative, contribution to the field of engineering, and bring about a dramatic change for the better due to women's abilities and advantages. A woman engineer can not only be better than a male engineer – she can also have a great positive impact on her male colleagues and bring about fundamental changes for the betterment in the engineering paradigm with its not-always-positive traits, habits, culture, mores, and characteristics. One of the most important changes can be mental – in the self-perception of the engineers, in the joy of creation, and in the enjoyment of engaging in engineering.

Women have engineering potential, certainly no less than that of men, and probably more. There is no reason why female students should not flock to engineering schools and succeed in their studies and later in their careers to a degree that not only does not decrease but even exceeds the degree of success of their male friends. In particular, female students can (and must) enjoy their studies even more than male students.

Despite the female advantages, engineering, and especially computer engineering, is a field in which women struggle to gain a foothold. Women make up barely a fifth in the field. Female students constitute a small minority of engineering students, and even among them the dropout percentage is significantly greater than the dropout percentage of male students. And even those who complete the degree are less inclined to continue working as engineers.

The explanations given are different. There is no unanimity regarding the fundamental causes. But there is a general agreement that whatever the original reasons may be, the immediate reasons for the minority of female engineers, including female engineering students, are that women do not enjoy occupation in the essential fields of engineering, are not connected to it, do not see themselves as an integral part of the world of engineering, feel alienated from the engineers, and from the students – their classmates. They are not integrated socially and mentally.

Like the chicken-and-egg paradox, we don't know if the different mental attitude of women, who are less self-confident as engineers and enjoy their studies less, is the cause of the objective difficulty and distance from the field, or vice versa.

The situation not only has not improved but has even worsened for many decades. Surprisingly and unfortunately, in the past the percentage of women in the field of computers was greater than today!

The way to a solution goes through a special reference to the different mental approach of female engineering students and an investment in providing the opportunity for female self-expression and realization in their special way. The objective aspect is of course important, but equally important is the immediate feeling, the state of mind, and the set of psychological factors including the student's perception of herself as an engineer, positive associations with the world of engineering and practice in engineering, enjoyment, and the feeling of confidence in her ability, motivation, and anticipation of her success.

One of the ways to achieve this goal can be the student's participation in an engineering maker lab in a place and in a way that will ensure the desired change. ERP of that kind could make use of knowledge in the field of professional and competitive sports, with its extraordinary empowerment of women and its supportive but highly challenging workout environment. Technology developed in sports, which has had enormous success in advancing women, can be applied in the field of engineering ERP.

In a special laboratory, operating in the spirit of Dewey, Montessori, Vygotsky, and Peper, a special atmosphere is created, more enjoyable and less stressful. On the other hand, from the very beginning of her studies, the woman student engages in challenging engineering projects and will feel that she can be, and is, successful. The goal in such a laboratory is young women engineers' personal and professional growth through intense R&D of project after project, at a higher level than in purely educational projects.

But on the other hand, the purpose of the projects is no longer limited solely to the acquisition of information or even knowledge; instead, emphasis is placed on the mental development of attitudes, beliefs, feelings, and ultimately shaping her personality as an engineer. All this is not achieved through lowering the standards, but on the contrary, as a result of more challenging projects, the student will be convinced that she can succeed in the most advanced subjects of engineering, and see the fruits of her labor materialize before her eyes.

In addition to the laboratory being a safe and pleasant place, with a feminine atmosphere, it teaches smart robotics and artificial intelligence by researching and developing a chain of projects that constantly increase in difficulty and challenge the student to the limit (and beyond). A female student has a higher motivation than men to engage in engineering (otherwise she wouldn't have succeeded in becoming a student). Therefore, she will be able to, gradually, during the 4 years of her studies, succeed in the most advanced projects, which are more difficult than usual, and closer to the work of an engineer in industry.

6.6 Ultra-orthodox woman computer engineering students

The ultra-orthodox women are, on the one hand, among the farthest from computer engineering undergraduate studies. They face great limitations due to their religious beliefs and are not encouraged (often actively discouraged) by their environment. They have significant gaps in their education, especially in STEM. Yet, on the other hand, it's one of the groups that needs computer engineering the most, for their personal self-realization, social mobility, overcoming their disadvantages and achieving success. This ERP is part of a special program tailored for ultra-orthodox women undergraduate engineering students. The program had great success with more than 90% graduating and then gainfully employed in the hi-tech industry.

6.7 Immigrant youth villages' engineering robotics playground

Immigrants have educational disadvantages for obvious reasons, such as language barriers, different educational backgrounds, and a lack of local networking. They especially need the mental and cognitive support that could come from ERP. Immigrants also have one significant advantage: greater motivation to study to break the glass ceiling. Naturally engineering and robotics would be the preferred field of study as they require fewer language skills and can have a greater impact. ERP is especially suited for that goal (Fig. 12).

Fig. 12: Presentation of the Youth Villages program results to Isaac "Bougie" Herzog, 11th President of Israel (first row third from the left), and members of Jewish Agency by the program head (first row, first on the left).

The program is delivered in the framework of seminars, workshops, and projects. The classes take place at the Knowledge Engineering and Robotics Laboratory at the Rupin Academic Center. Each meeting has theoretical and practical parts. Usually, each meeting can be divided into the following sections: theoretical part, hands-on activities related to construction (building different types of robots/artifacts), activities related to programming (problem-solving or writing solutions in code for a list of challenges), mini-project, presentation of the solutions, and conclusive summarizing competition.

The program has several main goals, from students acquiring essential twenty-first-century skills such as collaborative problem-solving, teamwork, communication, creativity, and critical thinking to increasing their confidence and knowledge in technology, robotics, programming, computer science, and artificial intelligence.

The following activities usually take place every year:
- several groups of pupils (up to 20 pupils) from different age groups (8–12th grade) attend the robotics activity every week, for 3 academic hours each session, during

a period of 10 weeks a year (2–3 months) during semester A. Each group will use robotic kits, 3D printer and other lab equipment.

– groups of pupils (up to 20 pupils) from different age groups (8–12th grade) attend the robotics activity every week for three academic hours each session, during a period of 8 weeks a year (2–3 months) during semester B. Each group will use robotic kits and other lab equipment.

Fig. 13: Team working based on Industrial ScorBot ER-4U arm robots systems project as close as possible to real-life industrial engineering.

The expected outcomes of the program:

1. Knowledge in the field of robotics and programming, such as building different robots for specific challenges and writing programs that solve this challenge with a help of robot/or robots.
2. Participants will acquire essential twenty-first-century skills such as collaborative problem-solving, teamwork, communication, presentation, and critical thinking.
3. Bringing together academia and youth villages. Bringing together students from the Faculty of Engineering (to serve as role models) and teens from a youth village (some from very low socioeconomic status, some new immigrants) (Figs. 13 and 14).

Hundreds of students have completed the different variations of this program. Over the years ERP has started to become a staple in immigrants' education.

Fig. 14: Team presents in exhibition/competition its very ambitious project of collective of EV3-LEGO Mindstorm robots cooperating in difficult environment.

6.8 ERP for minority school students

Minority students could have great obstacles to overcome, some objective (like fewer language skills, socioeconomic status of their parents, networking, and prejudice), some subjective (lack of confidence, and less identification with the engineering persona and culture of society in general). They will become great engineers as they have more reasons and motivation to work harder than others. They also inject vital diversity into engineering thought and culture. We have designed ERPs with these special characteristics in mind, aiming at all aspects, not just teaching the material. So, there was an effort made to build the ERP environment in a way that will enable it to use the project for more important goals such as language, confidence, integration, and long-term real-life scholar and industry outcomes (Fig. 15). Several groups in a variety of schools from primary to high school (100% minority student body) are using ERPs.

Fig. 15: Team using a variety of hardware to build a project, while sharpening their language, teamwork, and planning skills.

7 Conclusion

The ERP is the realization of pedagogical and psychological knowledge, dating back at least to the nineteenth century. We all agree on the need to foster the natural curiosity, support individual learning instead of frontal indiscriminate pontification, and develop ESs toward the creative productive wealth of our society. Until now, there was a lack of consistent well-defined practical infrastructure for this purpose. That changed with the technological advances, making ICT, computers and communications, electronics, and robotics not only desired as part of our life and education but also accessible due to professionals, resources, motivation, attitudes, and last but certainly not least, accessibility and affordability. The ERP was built as a framework for making to smooth, continuous evolution of engineering for everyone in our society.

The ERPs should start as early as possible, from the simplest making projects and evolve into R&D at the higher levels. It should be available to parents from birth to kindergartens, primary schools, middle schools, high schools, colleges, and universities, and even after the traditional educational age, it should be available to engineers, makers, and all others including organizations, companies, and employers. That will not only greatly improve the lives and happiness of individuals, and the wealth of our societies; it will create stronger netizens and empower communities.

Several directions are planned for investigation in future research. Long-term studies will be conducted to assess the impact of ERP programs on students' academic performance, creativity, and career choices, with the aim of determining the effectiveness of ERPs in fostering lifelong learning and ESs. Additionally, the integration of ERPs into existing curricula at various educational levels will be investigated, and the effects of ERP integration on student engagement and performance in traditional subjects will be examined. Efforts will be made to explore ways to seamlessly incorporate ERP activities into standard educational curricula and to develop standardized guidelines for embedding ERPs into both traditional and non-traditional educational settings.

References

[1] Tovani C. Do I really have to teach reading?: Content comprehension, grades 6–12. Routledge; 2023 Oct 10

[2] Chryssolouris G. Manufacturing systems: Theory and practice. Springer Science & Business Media; 2013 Mar 9

[3] Niemiec RM. Finding the Golden Mean: The Overuse, Underuse, and Optimal Use of Character Strengths. In: A second-wave positive psychology in counselling psychology. Routledge; 2023 Mar 31. pp. 183–201.

[4] Freidlin P, Littman-Ovadia H, Niemiec RM. Positive psychopathology: Social anxiety via character strengths underuse and overuse. Personality and Individual Differences, 2017 Apr 1, 108, 50–54.

[5] Montessori M. The montessori method. Transaction publishers; 2013 Nov 1

[6] Montessori M. The discovery of the child. Aakar books; 2004

[7] Papert S, Harel I. Situating constructionism. Constructionism, 1991 Apr, 36(2), 1–1.

[8] Papert S. The children's machine: Rethinking school in the age of the computer. New York; 1993.

[9] Vygotsky LS, Cole M. Mind in society: Development of higher psychological processes. Harvard university press; 1978

[10] Zviel-Girshin R, Rosenberg N. Montessori+ SE+ ICT= SEET: Using SE to Teach SE. In: Proceedings of the 5th international workshop on theory-oriented software engineering. 2016 May 14. pp. 1–7.

[11] Zviel-Girshin R, Rosenberg N. How to enhance creativity and Inquiry-Based Science Education in early childhood-robotic moon settlement project.

[12] Zviel-Girshin R, Luria A, Shaham C. Robotics as a tool to enhance technological thinking in early childhood. Journal of Science Education and Technology, 2020 Apr, 29(2), 294–302.

[13] Zviel-Girshin R, Rosenberg N. Educational Technology for pre-K Digizens. In: Proceedings of the 3rd international conference on information and education innovations. 2018 Jun 30. pp. 30–34.

[14] Dewey J. Experience and Education. In: The educational forum. Taylor & Francis Group; 1986 Sep 30. (vol. 50 No. 3 pp. 241–252).

[15] Gasco-Txabarri J, Zuazagoitia D. The sense of patterns and patterns in the senses. An approach to the sensory area of a Montessori preschool classroom. Education 3–13, 2023 Aug 18, 51(6), 979–987.

Ana Carolina Carius and Felipe de Oliveira Baldner

Maker movement and educational robotics in Brazil: from legislation to emerging practices in the school environment

Abstract: This work has, as its object of study, the implementation of the new high school (or secondary education) in Brazil, with a focus on the insertion of computational robotics, programming, and maker activities in each of the 26 states of the federation. As a general objective, we intend to evaluate the influence of the maker movement in the STEM area in Brazilian high schools, with a focus on the development of computational thinking in students, based on approaches in computational robotics. A mixed qualitative and quantitative study was carried out where, from a qualitative point of view, research was conducted on the state resolutions that guided the choice of the so-called learning paths and their implications in computational robotics. From a quantitative perspective, a study is presented based on data from the 2023 School Census, which aggregates all Brazilian schools. It was then verified the presence or absence of adequate infrastructure for the installation and operation of activities involving computational robotics in Brazilian schools. It is concluded that there are innovative initiatives aimed at expanding activities involving computational robotics at the high school level in Brazilian schools, although there is still a lack of significant investment in school infrastructure to accommodate this new moment in education in technology and innovation in Brazil.

Keywords: New high school, robotics, programming

1 Introduction

In 2018, the new secondary education was approved in Brazil [1], with a 4-year deadline for its implementation, that is, until the end of 2022. Because of this, the structure of Brazilian secondary education has undergone a substantial change. Until 2018, there was no possibility for students enrolled in this program to choose to delve deeper into topics in which they were most interested. The curriculum was the same for all students, regardless of their interests or the Brazilian state in which they were enrolled. It is worth noting that, in Brazil, the Education Guidelines and Bases Law (LDBEN, Lei de Diretrizes e Bases da Educação in Portuguese) [2] provides that sec-

Corresponding author: Ana Carolina Carius, Universidade Católica de Petrópolis, Petrópolis, Brazil, e-mail: ana.carius@ucp.br
Felipe de Oliveira Baldner, Universidade Católica de Petrópolis, Brazil, e-mail: felipe.baldner@ucp.br

https://doi.org/10.1515/9783111352695-003

ondary education, the object of study of this work, is the responsibility of the states, of which there are 26 in the Brazilian federative system. In other words, if the student were enrolled in high school in any of the 26 Brazilian states, they would have the same set of content, for any area of knowledge, at the same stage of schooling.

The rigid nature of secondary education and its low relevance to everyday problems experienced by Brazilian youth, such as social vulnerability, entry into the job market, or the choice of a higher education course, has made secondary education unattractive, especially for young people from families with lower incomes. This situation causes these young people to give up their studies to enter the job market prematurely or join organized crime, a significant problem in large Brazilian cities such as Rio de Janeiro and São Paulo [3].

Considering the weaknesses highlighted by the secondary education model in force until 2022, the Brazilian Ministry of Education and Culture (MEC, *Ministério da Educação e Cultura* in Portuguese) proposes a secondary education model that combines three specific characteristics: basic general training, a mandatory flexible part, and learning paths [4, 5]. Basic general training is responsible for offering a minimum and universal content for each of the areas of knowledge foreseen by LDBEN [2] for secondary education. The mandatory flexible part is composed of content that must be taught during high school, without specifying when or which curricular component will receive it. In this part, there is knowledge of digital culture, computational thinking, and life project (responsible for discussing future possibilities for young people), for example [6–8]. Finally, the learning paths are divided by area of knowledge, based on the training axes established by the National Common Curricular Base (NCCB or BNCC, *Base Nacional Comum Curricular* in Portuguese) [9]. These axes are languages and their technologies, mathematics and their technologies, natural sciences and their technologies, and human sciences and their technologies [10–12].

Based on the learning paths responsible for deepening knowledge in Mathematics and its Technologies, the object of study of this work is the insertions in the learning paths throughout the high school curricula of each of the 26 Brazilian states on topics such as robotics, programming, and digital technologies [13–16]. Therefore, this work aims to answer the following research question: regarding computational thinking and the STEM approach, how do school units perceive the need to include educational robotics and programming in the context of new high school?

Based on the guiding question of the research, the general objective of this work is to evaluate the influence of the maker movement in the STEM area, in Brazilian high schools, with a focus on the development of computational thinking in students, based on approaches in computational robotics and programming. As specific objectives, we can identify:

- the curricular changes present in high schools participating in the research focus on approaches that value computational thinking.
- analyze proposals specifically aimed at the STEM area and how such proposals are explained in the curriculum planning.

– evaluate the adoption (if any) of practices in educational robotics, verifying their relevance to the adopting school environment and the level of innovation of the proposal.

To carry out the intended study, achieve the proposed objectives, and answer the guiding question, a mixed qualitative and quantitative study was chosen. The study was characterized as qualitative because a documentary analysis was carried out of the 26 state resolutions that forwarded the Curricular Guidelines for New Secondary Education in Brazilian states, under the focus of learning paths offerings focused on robotics and programming. From a quantitative point of view, a survey was carried out, based on the 2023 School Census, of the physical and digital infrastructure for carrying out activities involving robotics and programming in Brazilian schools [17–19].

This chapter is structured as follows: Section 2 presents a quantitative study regarding the infrastructure of Brazilian schools and their appropriate facilities for educational robotics classes, based on the 2023 School Census, which is the main official report on the topic in Brazil. A qualitative study is presented in Section 3, in which documents from 26 Brazilian states regarding secondary education were analyzed, comparing the different proposals. In Section 4, an educational robotics proposal for high school students is presented, based on Computational Thinking. Finally, conclusions are presented in Section 5.

2 Quantitative study: the digital technology infrastructure of Brazilian schools

The School Census is a survey carried out annually by NIESR (National Institute of Educational Studies and Research Anísio Teixeira), which presents more than 300 specific pieces of information about each of the Brazilian schools, whether public (subsidized by the government) or private. Among the various pieces of information requested in the School Census, some refer to the digital technology infrastructure present in schools as well as internet access.

During the COVID-19 pandemic, one of the biggest challenges experienced by Brazilian schools was the lack of basic technological infrastructure to carry out school activities remotely [20–23]. Furthermore, many teachers and students cannot be considered digitally literate, as they do not have the minimum skills and competencies necessary for their adequate inclusion in the digital universe [24]. Therefore, analyzing aspects of digital inclusion in schools, from the perspective of investment in digital technological tools for carrying out pedagogical activities in Brazilian schools, is essential to understand the context in which we intend to discuss the adoption of pedagogical practices involving computational robotics and programming.

In the OECD (Organization for Economic Co-operation and Development) report "Education at a Glance" for 2023 [25], the discussion about the need to value initiatives that encourage young people toward so-called STEM careers is present, as had been addressed in the same report in previous years. Particularly noteworthy is the need for initiatives in this direction in developing countries, such as Brazil, as well as in specific minorities of the population that are not present in this area such as women [26, 27]. For the incentives referenced by these reports to occur, investments in technological and digital infrastructure are necessary in the Brazilian educational system.

According to the 2023 school census, Brazil has 180,230 basic education institutions, ranging from institutions that are exclusively for early childhood education, called early childhood education centers, to schools that only offer elementary education, schools that offer primary and secondary education, and schools that offer only secondary education. Of this, 70.8% of schools are public, while 29.2% are private and of this, 26.3% offer secondary education, as shown in Fig. 1.

Observing the data presented in Fig. 1, it is possible to verify that the computer laboratory, considered an environment for the use of computers at the beginning of the twenty-first century in Brazilian schools, is no longer a unanimous environment in schools. It is worth noting that, in 1997, the Brazilian federal government's ProInfo Program required the installation of a Computer Laboratory in each Brazilian public school [28]. It was not possible to install a Computer Laboratory in every Brazilian school, and, in addition, the equipment became obsolete due to the lack of public policies that could guarantee the functioning of the equipment [20].

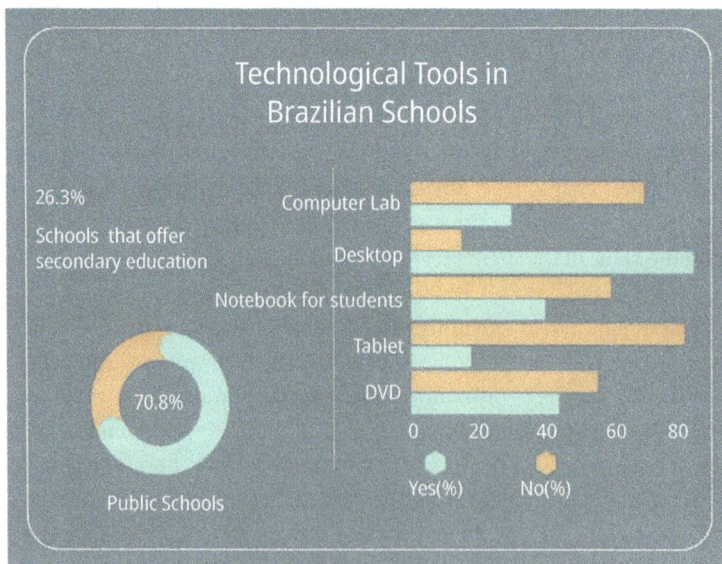

Fig. 1: Technological tools in Brazilian schools.

Regarding the absence of computer desktops in schools, reported in 15.3% of institutions, it appears that this percentage refers to some type of computer desktop present in the school and may even be used in daily management or secretarial school tasks. Therefore, 15.3% of Brazilian schools do not have any computers to carry out their daily tasks, regardless of the pedagogical function.

Considering portable computers, or laptops, it is observed that 40.2% of schools have this equipment available for student use, while tablets for student use are available in only 18% of institutions. The last observation refers to DVD players, also a disused audiovisual technology that is still available in 44% of Brazilian schools. The presence of obsolete technologies in 2023 in school institutions demonstrates the Brazilian educational system's delay in updating itself in the face of technological innovations available in students' daily lives.

However, as the research object of this study is the insertion of educational robotics in the context of Brazilian schools, we sought, in the public data offered by the 2023 School Census, some information regarding the presence of a Maker Laboratory in school institutions or the acquisition of instruments that refer to activities focused on robotics, such as Arduino boards, sensors, breadboards, and jumpers. This fact demonstrates the low relevance of robotics in the context of Brazilian schools, which leads us to the hypothesis that initiatives that include robotics and programming in Brazilian schools are still shy and isolated processes, not constituting a national public policy.

Another important issue regarding the adoption of pedagogical practices involving robotics and programming is the availability of Internet access in Brazilian schools. Fig. 2 shows the availability of the Internet in Brazilian schools: 91% of these have access to the Internet, with 40.5% of Brazilian schools allowing students to access the Internet and 63.3% of these using the Internet for pedagogical purposes.

From the data obtained from the 2023 School Census, it is possible to conclude that there is no way to make statistical inferences regarding the insertion in Brazilian schools of pedagogical practices that deal with computational robotics and programming. Furthermore, it is possible to notice that there is still a lack of internet and technological tools in some schools, which would make it impossible for Brazilian schools to adhere more to pedagogical practices that involve computational robotics and programming.

3 Qualitative study: analysis of secondary school curriculum documents

For qualitative studies, this work carried out the reading and analysis of the 26 curricular proposals presented by each of the Brazilian states, with a focus on including computational robotics and programming in the learning paths (or deepening paths) for young people who are attending the new high school in Brazil.

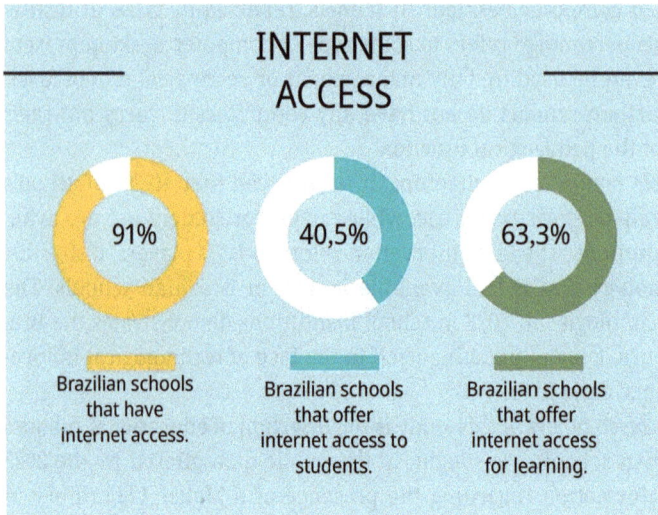

Fig. 2: Internet access in Brazilian schools.

In this study the learning paths planned for mathematics and its technologies were divided into five large groups: robotics, computer programming, financial mathematics, mathematics and society, and other topics that appeared less frequently. Fig. 3 summarizes the analysis of the proposals from the 26 Brazilian states.

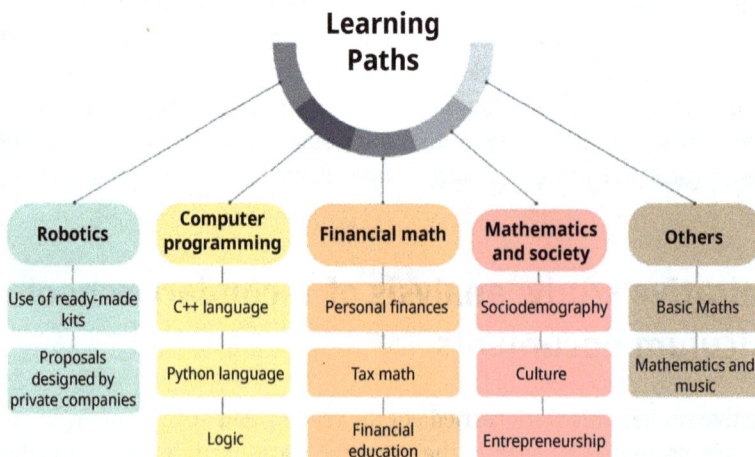

Fig. 3: Learning paths in Brazilian secondary schools.

Robotics learning paths are included in the high school curricula of 7 of the 26 Brazilian states. The proposals are varied and, in general, involve a partnership with a private company responsible for creating the curriculum proposal, supplying teaching materials, and training teachers. It is noteworthy that there was no proposal based on open-source software or hardware nor one that valued the construction of prototypes autonomously by students. Another issue that did not appear in any of the proposed documents was 3D modeling or more robust knowledge in electronics, confirming a curriculum with robotics insertions in a predefined way by a company that built all the teaching material in use.

Closely related to robotics, programming languages are not mandatory topics in Brazilian high schools, although computational thinking is present in the NCCB and must make up the mandatory flexible part of high school, which is not the object of study in this research. Once again, seven Brazilian states offer computer programming as a study option for high school students. Importantly, four states offer computational programming and computational robotics simultaneously in their learning paths, while others only offer computational robotics or computational programming. The presence of C++ and Python as programming languages, specifically, as well as the R language, which has a syntax closer to statistics, stands out. One of the states offers the discipline of logic, without specifying a computational language.

Although it is not the subject of study in this work, it is worth mentioning the significant number of curricular proposals that offer studies in financial mathematics. Six states contemplate these proposals, and none of them offer learning paths related to robotics and programming. It is important to relate this presence to the NCCB, which provides for financial education themes throughout basic education.

With important participation in the curricular proposals studied, there is an attempt to make mathematics less abstract by relating it to the daily lives of students, as also foreseen by the NCCB. In this way, the introduction of sociodemographic themes, including situations experienced in the job market and further closing the gap between mathematics and everyday activities, appears in the curricula of six states. However, in none of these proposals is there a relationship with robotics, nor is there any intersection with basic or physical electronics, to guarantee the insertion of pedagogical practices focused on computational robotics in Brazilian high schools.

It can be concluded from the qualitative analysis that there are formal attempts to include robotics and programming in the new secondary education. However, pedagogical practices are still diffuse, inspired by different objectives depending on the state of the federation to which the proposal is aligned. Furthermore, no proposal involves robotics for solving everyday problems, encouraging scientific investigation, and the creation of technological solutions that add value to the community to which the school belongs.

4 The future for secondary education in Brazil with robotics and computational thinking

After a thorough statistical evaluation of the Brazilian 2023 School Census, it is important to acknowledge the lack of data regarding the technological implementation of laboratories for the study of robotics and computational thinking.

The works of Andriola [29] and Moraes, Duran, and Bittencourt [30] list general skills a student must acquire when studying Robotics and Computational Thinking, which can be summarized in four main areas: (i) logical thinking; (ii) organization; (iii) problem-solving; and (iv) teamwork. These skills also help develop a creative mind and enhance the learning of math, physics, and languages (both Portuguese and English), characterizing this as a multidisciplinary subject. In a more specific approach, the students will learn about robotics, programming, and electronics.

The materials involved are diverse, ranging from Lego Mindstorms to the Arduino [17], as well as commercial kits, such as educational prosthetics [16] and even electronic waste [29]. While all these approaches can be used to develop the four main areas summarized above, the student is either stuck within the limits of their hardware, be it a Lego set or a Robotic Arm kit, or stuck behind a computer screen coding. What these approaches lack is a way to channel the teenagers' creativity and have them create their own robots.

One important aspect of using an Arduino board is the fact that it is an open-source project, meaning fewer constraints. Following this philosophy, the Otto DIY project enables users to build their projects from a base that is already established and lets their creativity guide them to solve their problems. The original project is open-source and encourages users to share their designs so others can use and change them and imbue them with their own personal touches.

A maker-style laboratory is needed to give this experience to students. This laboratory could be as simple as a 3D printer, several Arduino boards, sensors and actuators, and hand tools.

With these ideas in mind, a first-year, four-bimester course would be divided into 3D modeling, electronics and circuits, programming, and a final part for intersecting all the previous topics and making a functional robot. The first bimester would deal with the basics of 3D modeling and would use the Tinkercad platform. This choice was made considering that any of the conventional 3D modeling tools can be overwhelming for a teenager just entering high school. Tinkercad, on the other hand, is entirely intuitive, and students can start to apply their knowledge of spatial geometry, transforming basic shapes into complex projects that can be 3D printed. The basics of measurement and technical drawing can also be presented to the student so that they can 3D model a physical object. After this basic knowledge, the student should be able to understand how to translate a physical object into something that can be constructed by a 3D printer, with proper usage of measuring tools.

The second bimester would start by giving the student an introduction to algorithms. This will give the student the necessary knowledge to create sets of steps to solve a problem, for example, how to change a lightbulb. However, since the course isn't strictly for computer programming, as soon as the concept of loops is established, the subject of electronics should be readily introduced. With that approach, the student will be able to program something on the computer that will interact with real hardware.

The student would start with the classic "Hello World" example by making an LED blink. That would teach about digital outputs and timing. Afterward, the LED could be dimmed, introducing PWM (Pulse Width Modulation) as a form of controlling voltage. This will go on to sensors, such as an ultrasound device, and actuators, such as servomotors. In this way, the student will learn how both subjects interact with each other, creating interdisciplinarity between them. This would last for the entirety of the second and third bimesters.

For the last bimester, students would unite these three areas and create a 3D model that would be able to house all the electronics as well as program the onboard devices to solve a problem, such as making a self-driving two-wheel Otto robot that would navigate through a maze.

As previously stated, the tools used, whether hardware or software, should follow the open-source philosophy. The hardware used would be the Arduino, due to its widespread availability, both commercially and online, with a large quantity of libraries and development tools available. To avoid scaring the student with the rigidity of the C++ variant used by the default Arduino library, a more visual form of programming should be used. Blockly is a visual, open-source programming editor that uses blocks to perform actions and loops, and it can be used with many programming languages to generate code for the Arduino. Otto Blockly is built on top of this editor, providing blocks specifically for use with typical Otto hardware such as servomotors, buzzers, and ultrasound sensors.

To enable the teaching of the basics of the three areas, Autodesk's Tinkercad provides a simple solution for 3D modeling, programming, and electronics. While not an open-source project, it is free for use by anyone, student or not. Due to its small learning curve, Tinkercad was chosen instead of other open-source projects such as FreeCAD, which is aimed more toward engineering design.

As for the physical layout of the laboratory the students will use, they should have plenty of space on their workbenches for computers, 3D-printed parts, tools, and electronic devices. An open space would invite not competition between students, but collaboration and teamwork, as can be seen in the virtual rendering of Fig. 4 and the real laboratory of Fig. 5.

The proposed organization is for a 1-year course for any student who has never had any prior robotics, programming, or electronics knowledge. As the student goes through secondary education, their following years of robotics and computational thinking can be improved by starting to introduce new hardware in electronics such as ESP32, using C++ to program the Arduino, using full-fledged CAD software for 3D modeling, and even introducing artificial intelligence concepts.

Fig. 4: Digital rendering of a robotics laboratory for middle school.

Fig. 5: Real photograph of the robotics laboratory for middle school.

5 Conclusions

This work studied the insertion of computational robotics in the Brazilian educational system, based on the learning paths foreseen in the new Brazilian secondary education. Following the change in legislation for secondary education, which came into force in 2022, a qualitative study was carried out on the curricular proposals for secondary education in the 26 Brazilian states and a quantitative study of the infrastructure present in Brazilian schools, both under the proposed framework.

Considering the proposed research question: about computational thinking and the STEM approach, how do school units perceive the need to include educational robotics and programming in the context of new high school? It can be inferred that the perception of the need to include educational robotics and programming in the new secondary education is not unanimous, since only 10 states out of 26 (38.4%) of Brazilian states formalize this knowledge in their learning paths. The justification for this partial insertion is the heterogeneity of Brazil, with its regional specificities, so that each state has the autonomy to build its curricular proposal focused on the regional job market, the culture of the region, the needs of the young people who attend its schools, and reflecting on the profile expected of high school graduates in each state.

An influence, albeit small, of the maker movement can be seen in the construction of the proposals and, above all, based on the quantitative analysis of data relating to the infrastructure of Brazilian schools. By neglecting the information in the Brazilian School Census about the presence of a maker laboratory in schools, it is observed that there is still no public policy that values this space in Brazilian schools. Despite this, it is possible to identify the presence of this type of environment in several schools, highlighting the network of public schools in the states of Rio de Janeiro and São Paulo as an example.

However, pedagogical practices involving educational robotics are still independent and heterogeneous, highlighting the absence of proposals that validate open codes, 3D modeling, and robots aimed at the production of technology and innovation. These practices develop not only proper knowledge of robotics and programming but also logical thinking, problem-solving, teamwork, and creativity, alongside the interdisciplinary aspect involving regular subjects such as math, physics, and languages as well as opening the possibility of interactions with other subjects. In this sense, there is still a lot to advance in Brazilian schools as well as expanding, through public policies, maker activities in Brazilian high schools. This adaptation is fundamental for today's society, considering the symbiosis between Artificial Intelligence and everyday life that is envisioned soon.

Based on the responses presented to the proposed problem, future work highlights the need for on-site observations, such as case studies, on the implementation and development of secondary education proposals in the 26 Brazilian states since legislation on the topic is recent. Furthermore, there are ongoing changes in Brazilian high schools regarding the workload of mandatory subjects. In this sense, there may be changes from 2025 onward, which deserve due attention in future studies.

References

[1]   Brasil. Resolução n°3, de 21 de novembro de 2018. Brasília; 2018.

[2]   Brasil. Lei 9394 de 20 de dezembro de 1996. Ministério da Educação, Brasília, 1996, 34.

[3]   Filho Silva RB, Araújo RMDL. Evasão e abandono escolar na educação básica no Brasil: fatores, causas e possíveis consequências. Educação por Escrito, 2017, 8(1), doi: 10.15448/2179-8435.2017.1.24527.

[4]   Süssekind ML. A BNCC e o 'novo' Ensino Médio: reformas arrogantes, indolentes e malévolas. Retratos da Escola, 2019, 13(25), doi: 10.22420/rde.v13i25.980.

[5]   Cássio F, Goulart DC. A implementação do Novo Ensino Médio nos estados: das promessas da reforma ao ensino médio nem-nem. Revista Retratos da Escola, 2022, 16(35), 285–293.

[6]   de Souza GF, Lopes PTC. Aplicação do Pensamento Computacional no ensino, uma revisão sistemática de literatura. Interfaces Científicas – Educação, 2023, 12(1), doi: 10.17564/2316-3828.2023v12n1p144-165.

[7]   Rege A, Salgado LCC, Viterbo J. Pensamento Computacional no Contexto da Educação Brasileira: um Mapeamento Sistemático da Literatura nos diferentes Níveis de Ensino. Revista Novas Tecnologias na Educação, 2023, 21(2), Available https://orcid.org/0009-0008-8214-5110

[8]   Machado KK, Dutra A. Pensamento computacional: Uma análise da ementa do componente curricular no novo Ensino Médio. Ensino e Tecnologia em Revista, 2023, 7(2), Available http://periodicos.utfpr.edu.br/etr

[9]   Brasil, Base Nacional Comum Curricular, Brasília. 2018. [Online]. Available: http://download.inep.gov.br/acoes_internacionais/pisa/

[10] de Jesus DS, de Almeida MT, Wartha EJ. As transformações curriculares do Novo Ensino Médio em Sergipe: um olhar sobre a argumentação e o STEAM nas trilhas de aprendizagem. Com a Palavra, o Professor, 2023, 8(21), https://doi.org/10.23864/cpp.v8i21.957.

[11] Filho MHA, Welchen D, Socha MF, Ceretta ECM. A produção audiovisual como objeto de ensino e aprendizagem para as ciências humanas: experiências em trilhas de aprofundamento do novo ensino médio. Processando o Saber; 2023. pp. 15.

[12] Carius AC, Baldner FO, Maiworm AP. Robótica educacional no contexto do novo Ensino Médio: uma aplicação de código aberto. Revista Inter eEduca, 2023, 5(3), doi: 10.53660/rie.232.201.

[13] Fagundes CAN, Pompermayer EM, Basso MVA, Jardim RF. Aprendendo Matemática com Robótica. Revista Novas Tecnologias na Educação, 2005, 3(2), https://doi.org/10.22456/1679-1916.13943.

[14] Barbosa FC, Alexandre ML, Alves DB, Menezes DC, Campos GL, Nakamura YSN, Junior AJS, Lopes CR. Robótica Educacional em Prol do Ensino de Matemática. Sociedade Brasileira de Computação, 2015, doi: 10.5753/cbie.wie.2015.271.

[15] Vallim MBR, Herden A, Gallo R, Cardoso LR, Bitencourt LC. Incentivando carreiras na área tecnológica através da robótica educacional. In: Cobenge. Recife: Secretaria Executiva: Factos Eventos; 2009. pp. 1–10.

[16] Alves AL, Littike KA, Segatto BR, Moscon OS, Pessoa MS, Loyola GV, Botelho T. Utilização de kits de robótica como atividades práticas no ensino médio. Revista Caribeña de Ciencias Sociales, 2023, 12(8), doi: 10.55905/rcssv12n8-013.

[17] Kalil F, Hernandez H, Antunez F, Oliveira K, Ferronato N, dos Santos MR. Promovendo a robótica educacional para estudantes do ensino médio público do Brasil. In: Nuevas Ideas en Informática Educativa TISE. Porto Alegre: Sociedade Brasileira de Computação - SBC; 2013. pp. 1–4.

[18]   Ribeiro C, Coutinho C, Costa MF. A Robótica Educativa como Ferramenta Pedagógica na Resolução de Problemas de Matemática no Ensino Básico. In: CISTI. 2011. pp. 440–445.

[19]   Fraccanabbia N, Luvisa A, Bavaresco D. Planejamento de trajetórias polinomiais para robótica com Arduino Polynomial trajectory planning for robotics with Arduino. Revista Eletrônica da Matemática, 2018, 4(1), https://doi.org/10.35819/remat2018v4i1id2758.

[20]   Carius AC. Brazilian Public Schools and COVID-19: A picture of an asynchrony between technology and teaching practices. Conjecturas, 2022, 22(1), doi: 10.53660/conj-507-704.

[21]   Carius AC, Oliveira MA. Relação família-escola em tempos de COVID-19: discutindo questões cotidianas. Humanidades & Inovação, 2022, 9(6), 164–180.

[22]   Saraiva K, Traversini C, Lockmann K. A educação em tempos de COVID-19: ensino remoto e exaustão docente.Praxis Educativa, 2020, 15. doi: 10.5212/praxeduc.v.15.16289.094.

[23]   Carius AC. Democratic Management in Times of COVID-19: Perspectives and Challenges for Brazilian Public Schools. International Journal for Innovation Education and Research, 2021, 9(8), https://doi.org/10.31686/ijier.vol9.iss8.3288.

[24]   Carius AC, Scartoni FR. The Teaching Process in times of COVID-19: Between the possible and the idealized. Revista Brasileira de Política e Administração da Educação, 2023, 39, doi: 10.21573/vol39n12023.129631.

[25]   Education at a Glance 2023. in Education at a Glance. OECD, 2023. doi: 10.1787/e13bef63-en.

[26]   María AB, Estébanez E. Uma equação desequilibrada: aumentar a participação das mulheres na STEM na LAC. 2022. Available: www.unesco.org/open-access/

[27]   Lorenzoni MJ, Beserra K, Bezerra L, Jurgina LQ, Júnior LSR. Gurias de Comp: A permanência de mulheres em cursos de graduação em Ciência e Engenharia da Computação. Communications and Innovations Gazette, 2023, 1(1), 96–102.

[28]   Carius AC. Pedagogia Digital. 1st ed. Curitiba: CRV; 2023.

[29]   Andriola WB. Robótica Educacional em Escolas Públicas do Ceará: avaliação dos impactos sobre o desenvolvimento de competências discentes e a qualidade do ensino. Revista Docentes, 2024, 9(25), 59–72.

[30]   Moraes JPA, Duran RS, Bittencourt RA. Robótica Educacional E Habilidades Do Século XXI: Um Estudo de Caso com Estudantes do Ensino Médio. In: EduComp'23. Porto Alegre: Sociedade Brasileira de Computação: 2023. pp. 173–183. https://doi.org/10.5753/educomp.2023.228195

Nerea López-Bouzas, Jonathan Castañeda Fernández, and M. Esther del Moral Pérez

Robotics and computational language in the school setting: a review of research

Abstract: The aim of this chapter is to provide an overview of research published between 2017 and 2024 on the use of robotics in the school environment (aged 3–12). By narrowing the search according to these criteria and adopting the principles of the PRISMA statement, 44 studies were found, 18 from early childhood education interventions and 26 from primary education. These were subsequently classified according to the purpose for which robots are incorporated into classrooms: (1) analysis of teachers' opinions on the implications of this task as well as the assessment of their digital competence; (2) interventions to stimulate communicative competence; emotional learning and students' socio-emotional skills using social robots or to enhance their computational thinking. The results show that most aim to develop students' computational language and focus on prototype testing. Some studies indicate that robots increase motivation toward learning, while others highlight their limitations in improving students' socio-emotional skills as they do not recognize emotions. However, they do promote autonomy in certain tasks. Finally, few studies analyze teachers' digital competence to incorporate these resources into the classroom. In conclusion, to successfully replicate these interventions, studies need to describe the evaluation instruments applied. Additionally, sample sizes should be expanded, and it would be desirable to create a network of researchers around this topic to develop effective intervention models.

Keywords: Robotics, computational language, school, research review

1 Introduction

Currently, the Horizon Report, Teaching and Learning Edition [1], and the ICT Competency Framework for Teachers [2] advocate for enhancing the digital competencies of both teachers and students, as these skills are essential for interacting in a highly technological world. The logic of computational thinking (CT) is becoming necessary

---

Corresponding author: Nerea López-Bouzas, Department of Education Sciences, University of Oviedo, Oviedo, Spain, e-mail: lopeznerea@uniovi.es

Jonathan Castañeda Fernández: Department of Education Sciences, University of Oviedo, Spain, e-mail: castanedajonathan@uniovi.es

M. Esther del Moral Pérez: Department of Education Sciences, University of Oviedo, Spain, e-mail: emoral@uniovi.es

https://doi.org/10.1515/9783111352695-004

to interact with increasingly sophisticated machines, revolutionizing the teaching-learning process. Wing [3] considers CT a skill for systematically and logically solving problems, encompassing dimensions such as algorithmic thinking, generalization, abstraction, and evaluation. Additionally, Vuorikari et al. [4] identify it as a key component in stimulating digital competence. Clearly, teachers need guidelines to foster critical thinking from an understanding of the role these emerging technologies play in education. Furthermore, they must be capable of integrating them into the school curriculum to optimize the teaching-learning process, aiming to enhance their students' reflective and problem-solving abilities creatively and collaboratively.

Thus, educational robotics (ER) emerges as a methodological trend set to consolidate in the coming years. It combines engineering and programming with educational objectives to stimulate cognitive, social, and motor skills from an early age [5, 6]. The integration of ER in schools has emerged as an innovative pedagogical approach, leveraging technological advancements to promote learning in a playful and interactive manner [7]. Additionally, computational language is a key element for programming commands that a machine, whether a robot or another mechanical device, must execute [8]. Today, there is increasing human interaction with machines, necessitating an understanding of their language and adapting thinking to their cognitive structuring [9]. Consequently, educational contexts are seeing a proliferation of training experiences aimed at acquiring computational language and incorporating robots.

These innovative proposals presuppose specific teacher training. Educators must have a high level of digital competence not only to handle robots and/or know various programming languages [10, 11] but also to implement them appropriately in their classrooms, considering the context and the needs of their students. However, the introduction of robots and AI-supported devices in schools encounters some resistance from teachers, as not all are fully aware of their potential or have experimented with them [12]. The fear of the unknown and lack of qualification [13], along with the lack of financial resources to equip schools with these sophisticated educational materials [14] are significant factors limiting their real incorporation into schools. Consequently, initial studies have emerged to gauge teachers' opinions and perceptions regarding the usefulness of robotics, including its ethical implications [15], recognizing that teachers are the true drivers of classroom innovation.

The structure of the chapter is detailed below. In Section 1, a general overview of the current importance of using robotics in the educational field is provided. In Section 2, the evolution and role of robotics in early childhood and primary education are explored, highlighting various experiences that utilize these resources and specifying the skills stimulated by the interaction and programming of robots. In Section 3, the search criteria for the selection of studies are described, detailing the review process and the classification of the selected research. In Section 3.1, the number of studies found and their breakdown according to the objective of each is specified. In Section 4, an exhaustive analysis of the research is conducted, considering its subject matter. In Section 4.1, studies exploring teachers' opinions on the integration of robotics in the classroom and/or assessing their digital competence to use it are analyzed. In Section 4.2, the selected studies are exhaustively detailed according to their objectives: (a) stimulation of communicative competence, (b) stimulation of socio-emotional skills with social robots, (c) acquisition and development of computational thinking. In Section 4.3, a synthesis of the studies found is made, observing common points among them. In Section 5, the most significant findings of the research are interpreted, highlighting patterns, trends, and discrepancies. Practical and theoretical implications of integrating robotics in the classroom are reflected upon. Finally, in Section 6, the main results are summarized and recommendations for future research are proposed.

2 Robotics in the school context

The phenomenon of robotics is impacting classrooms, where there is receptive teaching staff willing to incorporate it. However, there is another sector resistant to it due to its perceived complexity and the lack of necessary training. Thus, research has emerged to understand opinions on the incorporation of robotics in the classroom, analyzing some of the experiences conducted [16, 17] and addressing the limitations teachers face in integrating robots with educational resources [18]. Their perception of how the use of robotics in the classroom influences the enhancement of their digital competence is also noted [19]. Undoubtedly, qualified teaching staff to utilize robotics for educational purposes are inclined toward its integration in classrooms in diverse and innovative ways, fostering the stimulation of various competencies in students.

On the other hand, companies that market such artifacts take advantage of testing their prototypes in some schools, seeking feedback to improve them. They solicit feedback from teachers to implement improvements in subsequent versions and address recurring issues encountered by students [20]. The advancement of this industry continually offers new features, allowing teachers to use robots as conversation aids in other languages, facilitating learning associated with new vocabulary, pronunciation, grammatical structures, etc. [21].

In this regard, research is emerging from these innovative experiences based on the implementation of robotics, both in early childhood education and primary education classrooms, where robots assume a leading role with various purposes, especially to promote social interaction or serve as tutors to assist students in tasks such as reading and social skills [22]. Additionally, the emergence of studies examining the extent to which the incorporation of robots is contributing to stimulating communicative competence at these levels of basic education, for vocabulary acquisition, pronunciation, etc., is noted [23]. Moreover, these artifacts are utilized as storytellers to enhance engagement, attention, and comprehension of stories [24].

Other research focuses on identifying the opportunities provided by so-called "social robots" to stimulate socio-emotional skills in early childhood education students through the use of conventional communication forms and the assimilation of social norms [25]. There are also experiences analyzing the ability of children to interact with these robots in playful contexts [26].

Furthermore, there are studies specifically focused on the acquisition of computational language at early ages, i.e., the assimilation of structured forms to interact with machines. Some research focuses on activating CT, considering the robot as a key actor in the development of STEAM proposals in primary education [27]. Some aim to activate reflective and metacognitive processes through activities involving following certain sequences to achieve a goal [28]. Others specifically analyze how robots help activate algorithmic thinking, based on response patterns that promote the resolution of logical-mathematical problems [29].

Moreover, increasingly intuitive programming languages have been created based on the use of blocks to configure actions, such as Arduino and Scratch, for example, to create animations; others serve to establish sequences of actions for a robot to move autonomously through a space, engage in humanized conversations, etc. The skills that can stimulate interaction with robots and computational programming are synthesized in Fig. 1.

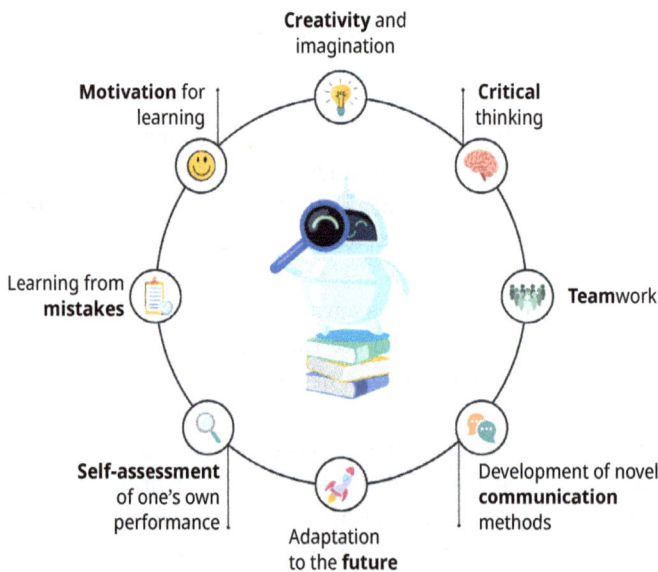

Fig. 1: Skills stimulated by interacting with robots and computational programming.
Source: Own elaboration.

Therefore, the aim of this review is to provide an overview of the most recent research on the use of robotics in the school environment. First, it focuses on teachers' opinions regarding the implications of its use and its contribution to enhancing their digital competence. Second, it examines the outcomes of interventions conducted with students aged 3–12 years, utilizing robots for various purposes. Some interventions are oriented toward stimulating communicative competence and socio-emotional skills, while others focus on the acquisition and development of CT.

3 Methodology

This systematic review adopts the principles of the PRISMA statement [30] and standardized methodological guidelines proposed for the preparation of high-quality systematic reviews [31]. A search was conducted on Google Scholar to select articles that meet the following criteria: (1) the use of robotics in the educational context (aged 3–12 years); (2) published between 2017 and February 2024 in the international context and in English; (3) studies available in databases such as SAGE, SCOPUS, SpringerLink, Web of Science, Wiley Online Library, Oxford Journals, Cambridge Journals, Google Scholar, and Science Direct. Systematic reviews were excluded. The keywords used for the search were: "robot/robotics + Early Childhood Education/Preschool Education/ Kindergarten" and "robot/robotics + Primary/Middle Education".

3.1 Procedure

The phases that have structured the search and selection of articles are described in Fig. 2.

Finally, the sample consists of 44 studies, 18 focusing on early childhood education and 26 on primary education. After careful reading, the studies were grouped according to the purpose or objective of incorporating robots into classrooms:

3.1.1 Studies focused on teachers

a) Analysis of teachers' opinions on the contributions of robots and/or computational language in their classrooms. Information on opportunities, limitations, impact, etc., as well as guidelines for designing future prototypes, addressing their deficiencies, is gathered through interviews and questionnaires [15, 16, 18, 20, 32].
b) Confirmation of their perception regarding the digital competence required to incorporate robotics in the classroom [17, 19].

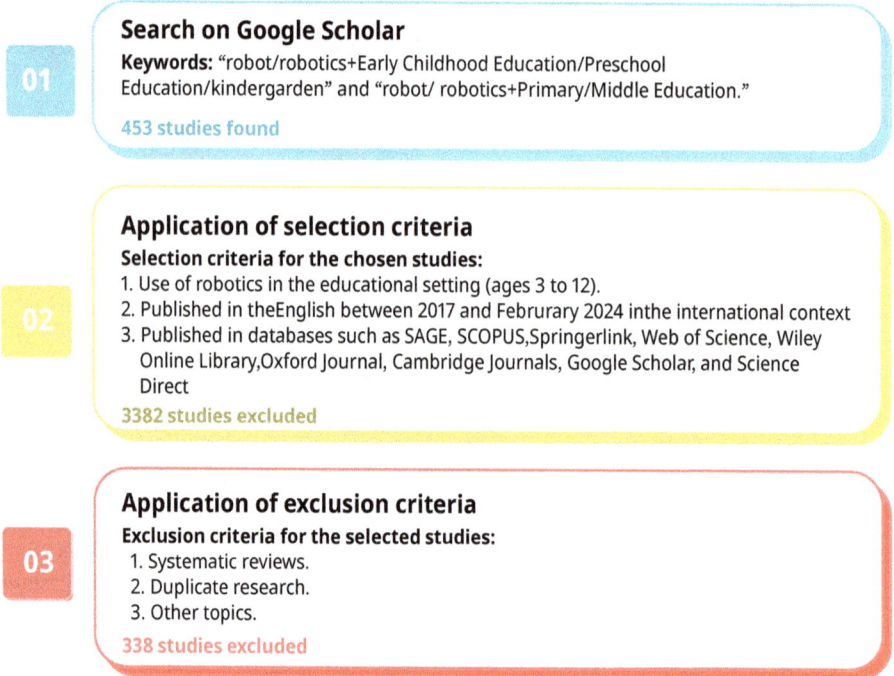

Search on Google Scholar
01
Keywords: "robot/robotics+Early Childhood Education/Preschool Education/kindergarden" and "robot/ robotics+Primary/Middle Education."

453 studies found

Application of selection criteria
02
Selection criteria for the chosen studies:
1. Use of robotics in the educational setting (ages 3 to 12).
2. Published in theEnglish between 2017 and Februrary 2024 inthe international context
3. Published in databases such as SAGE, SCOPUS,Springerlink, Web of Science, Wiley Online Library,Oxford Journal, Cambridge Journals, Google Scholar, and Science Direct

3382 studies excluded

Application of exclusion criteria
03
Exclusion criteria for the selected studies:
1. Systematic reviews.
2. Duplicate research.
3. Other topics.

338 studies excluded

Fig. 2: Phases of the search process.
Source: Own elaboration.

3.1.2 Studies focused on students

a) Studies aimed at analyzing how robots facilitate and enhance communicative competence, utilizing various strategies and methods that encourage interaction and verbal expression of students in their native or foreign language: (1) native language [24, 23–36]; (2) foreign language [37–40].

b) Studies aimed at exploring the contribution of social robots to improving emotional learning and stimulating socio-emotional skills: Baxter et al. [59], Heljakka et al. [60], Samuelsson [26], Serholt [61], Serholt [62], and Sung et al. [25].

c) Studies focused on analyzing how robot programming enhances skills associated with CT, combining robots with simple or graphic programming environments that allow students to transform them into intelligent objects and respond to external stimuli: (1) button-based programming [27, 28, 33, 41, 43–47] and (2) block-based programming [29, 48–56].

4 Results

Below is a comprehensive analysis of the research focusing on its object of study and purposes.

4.1 Studies focused on teachers

Table 1 summarizes the studies that analyze teachers' opinions on the contributions of robots and/or computational language in their classrooms as well as those that evaluate their digital competence. Specifically, it identifies the authorship, publication year, nationality, and educational level as well as the research design followed: the objective, the robot used, the methodology, the procedure, the instruments used to collect information, the sample, and the most relevant results.

Tab. 1: Teachers' opinions on robots and digital competence in teaching.

Authorship (date), nationality, stage	Research design		
	Robot objective	Method procedure and instrument (*N*)	Results
Cho et al. [32], USA, primary education	*Bee-Bot*® y *Cubelet* analyze the implications of educational robotics (7–8 years).	Qualitative: Three-case study. Video recording of interactions and surveys.	The results are positive. There is a demand for teacher training in this field.
Chootongchai et al. [21], Thailand, primary education	Understand how the design of a robot for educational purposes should be according to teachers.	Mixed: Qualitative interviews combined with quantitative surveys (*N* = 510).	Teachers propose six dimensions for designing an educational robot.
Gavrilas et al. [18], Greece, preschool education	Analyze teachers' opinions regarding the potential of robotics in their teaching tasks.	Quantitative: Attitude toward robotics potential, perspectives, and interest in this methodology are collected.	Its importance and interest in incorporating it are recognized, but teachers feel undertrained.
Janicki and Tenberge [19], Germany, primary education	*BlueBot* and *OzoBotare.* Analyze if teachers possess competencies to use robots.	Quantitative experimental: pre-test and post-test design. Three groups are compared: GE, GC, and reference group.	No results yet. It is an ongoing research with no current outcomes.

Tab. 1 (continued)

| Authorship (date), nationality, stage | Research design | | |
	Robot objective	Method procedure and instrument (N)	Results
Poolsawad et al. [16], Thailand, primary education	Analyze teachers' perceptions of the integration of robots in the classroom.	Mixed: A robot was created and tested (N = 20) based on teachers' guidelines (N = 510). Interviews and questionnaires were used.	The prototype improved students' skills and teaching strategies.
Sánchez-Rivas et al. [63], Spain, primary education	Evaluate teachers' digital competence: strategies and tools for teaching block-based programming (20–70 years).	Quantitative: The instrument used is a structured questionnaire assessing the level of digital competence of teachers (N = 300).	Teachers do not exhibit a good level of digital competence in robotics and programming areas.
Serholt et al. [15], Sweden, primary education	Explore teachers' opinions on the ethical implications of robots in the classroom.	Qualitative: Questions about privacy, subsequent effects, etc. were asked through focus groups (N = 77).	Teachers consider the implementation of robots useful, but not autonomous robots.

Source: Own elaboration.

As observed in Tab. 1, the objectives of these investigations are diverse. Janicki and Tenberge [19] and Sánchez-Rivas et al. [17] analyze whether teachers have the necessary digital competencies to use robots, emphasizing their deficiencies in programming and computational language. Others inquire about what they deem necessary to incorporate robotics in classrooms from methodological and ethical perspectives and their assessment regarding its formative potential. Poolsawad et al. [16] and Serholt et al. [15] highlight its benefits compared to conventional methods. Meanwhile, Cho et al. [32] and Gavrilas et al. [18] conclude that teachers are receptive to its incorporation but demand more training.

Chootongchai et al. [20] identify teachers' opinions on the requirements for designing robots for educational purposes and propose improving their appearance to make them more friendly, adopting humanoid facial expressions similar to those of an animal or animated doll, and integrating flashing lights, sound, vibration, or touch response, in order to achieve greater adaptation to students. They also emphasize the importance of robots presenting nonrepetitive responses and moving autonomously. Additionally, they suggest that robots can read and narrate stories, provide feedback to students, respond to specific commands, comment, communicate their feelings and emotions, etc.

4.2 Studies focused on students

4.2.1 Stimulation of communicative competence

Below are the studies that address the use of educational robotics to stimulate students' communicative competence in their native language (Tab. 2).

Tab. 2: Utilization of robots to stimulate communicative competence associated with the native language.

Authorship (date), nationality, stage	Research design		
	Robot objective	Método Procedimiento e instrumento (*N*)	*Robot* objective
Ioannou and Bratitsis [23], Macedonia, early childhood education	*Sphero SPRK*. Stimulate knowledge of animals (3–6 years).	Qualitative. Observation of responses to questions about animals (shape, size, etc.).	Students correctly identified the characteristics.
Michaelis and Mutlu [33], USA, primary education	*Minnie*: Design a robot to support home reading tasks (11–12 years).	Qualitative: Explore the habits and views of families (*N* = 8) about reading.	The robot helps them socially engage with reading.
Moreno and Rodríguez-Muñoz [34], Spain, early childhood education	*Super.Doc*: Stimulate linguistic competence (4–5 years)	Quantitative: Comparison of results (pretest/posttest) measured with the TEPI scale (*N* = 21).	Benefits are observed in favor of fostering meaningful learning of linguistic skills.
Robben et al. [24], early childhood education, Holland	Narrator robot to analyze if gender affects how children perceive narrators.	Qualitative: Observation of students' interaction with the narrator robot (*N* = 64).	Children did not anthropomorphize the robot when its gender matched theirs.
Westlund et al. [35], USA, early childhood education	Social robot: Analyze whether vocabulary learning is better supported by robots or peers (2–5 years).	Quantitative: Images of animals identified by a peer and a robot were shown, and then their retention was assessed (*N* = 36).	There were no differences in children's vocabulary retention with the support of the robot or peers.
Westlund et al. [36]., USA, early childhood education	Compare the effect of a facial expression narrator robot and another without it on vocabulary acquisition (3–6 years).	Quantitative: Concentration, engagement, story emulation, and vocabulary retention (*N* = 45) were evaluated with two robots.	An expressive robot leads to better results than one with a flat voice.

Source: Own elaboration.

Primarily, these studies focus on stimulating linguistic skills, specifically on vocabulary enhancement [23, 34, 35]. Specifically, Westlund et al. [36] compare the effectiveness of a storytelling robot with facial expression and another with no expression for vocabulary acquisition. Michaelis and Mutlu [33] utilize robots to support reading at home, emphasizing that students and their families improve their outcomes. And Robben et al. [24] analyze how the voice of a storytelling robot impacts children's perception and comprehension.

On the other hand, Tab. 3 identifies those studies that stimulate communicative competence associated with a foreign language.

Tab. 3: Utilization of robots to stimulate communicative competence associated with a foreign language.

Authorship (date), nationality, stage	Research design		
	Robot objective.	Método Procedimiento e instrumento (*N*)	*Robot* objective
Alemi et al. [37], Iran, early childhood education	*Assisted language learning (RALL)*: Stimulate motivation to learn English (3–6 years).	Qualitative: Observation of anxiety levels, interaction, and motivation based on intervention videos (*N* = 19).	Motivation was stimulated, and the children considered the robot their friend.
De Wit et al. [38], Netherlands, early childhood education	Acquire new vocabulary of a foreign language (4–6 years) with a robot.	Quantitative: Vocabulary activities were conducted using gestures.	Positive effect on the long-term memorization of new words.
Haas et al. [39], Netherlands, early childhood education	Stimulate learning a foreign language (3–6 years) with a robot.	Quantitative: Observation of children's interaction with the robot during activities (*N* = 65).	Students performed activities more independently with the assistance of the robot.
Rintjema et al. [40], Netherlands, early childhood education	Social robot to improve communication in a second language (3–6 years).	Qualitative: Observation of proficiency and motivation during communication in the second language (*N* = 15).	Improved language skills and motivation toward learning were observed.
Van den Berghe et al. [64], Netherlands, early childhood education	Analyze if learning a second language improves with the help of a robot (3–6 years).	Qualitative: Observation of the performance of activities alone, with peers, or with a robot to learn a second language.	Only the result of a language activity improved when done individually.

Source: Own elaboration.

Alemi et al. [37], Haas et al. [39], and Rintjema et al. [40] utilize robots to improve oral communication in English, while De Wit et al. [38] emphasize vocabulary enhancement in this second language. The studies analyzed use robots designed by the researchers to conduct interventions, while others integrate commercial robots: Assisted Language Learning (RALL) (https://www.alelo.com/rall-e-project/) for language learning; Super Doc (https://cutt.ly/9eraZJtK), to stimulate linguistic competence; and Sphero SPRK + (https://cutt.ly/xeraX33I), a robotic sphere used to expand animal vocabulary.

4.2.2 Stimulation of socio-emotional skills with social robots

Table 4 shows the research focused on the use of social robots to improve emotional learning and stimulate socio-emotional skills.

Tab. 4: Utilization of social robots to stimulate socio-emotional competence.

Authorship (date), nationality, stage	Research design		
	Robot objective	Método procedimiento e instrumento (*N*)	*Robot* objective
Baxter et al. (2018), United Kingdom, primary education	*Nao*: Evaluate the effect of personalization on social behavior (7–8 years).	Exploratory quantitative: Analysis of the impact of using a social robot compared to traditional methods (*N* = 59)	Personalization using a social robot enhances learning.
Heljakka et al. (2021), Finland, Early childhood education	Robot dog to stimulate socio-emotional skills (5–6 years).	Experimental qualitative: Observe the difference between interaction with a real dog and a robot dog (*N* = 16)	Socio-emotional skills improve and strategies for using the robot dog are outlined.
Samuelsson (2023), Sweden, early childhood education	*Blue-Bot*: Analyze how children play with the robot and their opinion on AI (1–5 years).	Qualitative: Ethnographic study (observation and interviews) over 7 months (*N* = 38)	Children comprehend robotics and AI and their current uses after interacting.
Serholt (2018), Sweden, primary education	*Pepper*: Explore the causes of errors in interaction with a robotic tutor (9–12 years).	Qualitative: Observation of videos where students interact with a robot (*N* = 6)	Forty-one indications of malfunction were observed.
Serholt (2019), Sweden, primary education	Explore children's retrospective perceptions of their relationship with a robot (9–12 years).	Mixed: Surveys and focus groups (*N* = 34)	Emphasize the lack of emotional recognition in robots.

Tab. 4 (continued)

Authorship (date), nationality, stage	Research design		
	Robot objective	Método procedimiento e instrumento (*N*)	*Robot* objective
Sung et al. [25], South Korea, early childhood education	*KIBO*: Determine if robotic kits influence students' cognitive and social abilities (5–6 years).	Quantitative, Quasi-experimental research: analyze results (pretest/posttest) between an experimental group using the robot and a control group that does not (*N* = 450).	Robots enhance the effectiveness of a STEAM program, stimulating various competencies.

Source: Own elaboration.

Samuelsson [26] evaluates students' perceptions while playing and interacting with a robot. Sung et al. [25] identify if robotics kits impact cognitive and social skills. Haas et al. [39] analyze the positive impact of personalizing robots on students' social behavior. On the other hand, Serholt [61] documents the problems derived from interaction with a robotic tutor to address them. The same author gathers students' perceptions while interacting with a robot, highlighting their lack of emotional recognition.

Some studies utilize the KIBO robot (https://kinderlabrobotics.com/kibo/), which allows assembling physical blocks to create sequences of commands that it scans to execute. Others opt to integrate Blue-Bot (https://cutt.ly/XeraVJwc) into their interventions, which include Bluetooth, allowing programming from mobile devices or tablets. Some use Nao (https://www.aldebaran.com/en/nao), designed for conversation, while others employ Pepper (https://cutt.ly/deraBCyT) to interact with people, detecting emotions for assistance purposes, or to teach programming, robotics, and artificial intelligence, promoting skills for social interaction.

4.2.3 Acquisition and development of computational thinking

Tables 5 and 6 compile the research addressing the use of educational robotics to stimulate CT. Subgroups were established based on the type of programming used either with buttons (Tab. 5) or with blocks (Tab. 6).

The objectives of the research specifically aimed at stimulating computational language are diverse. However, most of them focus on analyzing the acquisition and development of CT [41, 45, 55] and its implementation in STEAM proposals [27, 43]. Specifically, Michaelis and Mutlu [33] analyze the advantages of robotics in increasing interest in science, noting a better understanding of scientific processes. On the other

Tab. 5: Acquisition and development of computational language through button programming.

Authorship (date), nationality, stage	Research design		
	Robot objective	Método procedimiento e instrumento (N)	*Robot* objective
Caballero et al. [41], Spain, early childhood education	*Bee-Bot®*: Activate programming skills (3–6 years).	Quantitative: Problem-solving activities were carried out, and data were collected through questionnaires.	The students achieved good results in the scales measuring their abilities.
Chatzopoulos et al. [27], Greece, primary education	Stimulate computational thinking with robotics and STEAM.	Quantitative: Observation of students' performance during STEAM activities.	The robots activate their interest in programming.
Chen et al. [42], USA, primary education	Physical and virtual robots (10–11 years). Stimulate computational thinking.	Quantitative: Testing a physical and a virtual robot using pretest and posttest ($N = 767$).	The students increased their computational skills.
Ching et al. [43], USA, primary education	Stimulate mathematical skills with STEAM methodology and robots (9–12 years).	Mixed: Interviews, focus groups, and Likert scale questionnaires were combined ($N = 18$).	They improved their computational skills and teamwork.
Di Lieto et al. [44], Italy, early childhood education	*Bee-Bot®*: Stimulate executive skills (5–6 years).	Quantitative: Evaluation before, during, and after the intervention with a neuropsychological battery ($N = 12$).	Significant improvement in visuospatial working memory. Favors inhibition.
Jones and Castellano [28], UK, primary education	*Nao*: Stimulate reflective and metacognitive processes (10–12 years).	Quantitative: Activities to use a compass, read maps, and measure distances (pretest/posttest) ($N = 24$).	The results were positive. They improved self-control and seeking help.
Kopcha et al. [45], USA, primary education	*Danger Zone*: Stimulate computational thinking (10–11 years).	Qualitative:. Mathematical and programming activities. Response to an online survey and descriptive coding ($N = 263$).	The environment designed supported by robotics stimulates problem-solving.

Tab. 5 (continued)

Authorship (date), nationality, stage	Research design		
	Robot objective	Método procedimiento e instrumento (N)	*Robot* objective
Michaelis and Mutlu [65], USA, primary education	*Minnie*: Analyze the impact of a robot on increasing interest in sciences (10–12 years).	Mixed: Design of a social robot and testing during the reading of a science book (pretest/posttest) (N = 63).	They increased their situational interest and made more precise statements.
Sáez et al. [46], Spain, primary education	Evaluate the potential of educational robotics and visual block programming (10–11 years).	Quantitative: Quasi-experimental research, comparison of a group using robots and block programming with a control group (N = 107).	They acquire computational language by using robotics and visual block programming.
Timur et al. [47], Turkey, primary education	Examine students' perceptions of robotic coding (7–9 years).	Qualitative: Creation of drawings and oral explanations. Descriptive analysis (N = 14).	The children did not include themselves in the drawings but humanized the robot.

Source: Own elaboration.

hand, Miranda [51] evaluates the skills to program a robot to navigate and interact in the scenario of a classic story.

Hsiao et al. [49] enhance students' computational skills during the creation of their own robots. Similarly, other research aims to activate programming skills [48], algorithmic thinking [29], executive functions [44], or reflective and metacognitive processes with spatial activities [28]. Noh and Lee [52] analyze the impact of a programming course to acquire programming skills, and Sáez et al. [46, 53] evaluate the possibilities of block programming.

Others choose to analyze students' perceptions of robotics and computational language [56], and likewise, Sullivan and Bers [54] study students' perception of the implications of educational robotics and conclude that it is positive, noting an increase in girls' interest in technology. Additionally, Martínez et al. [50] analyze whether the use of the robot is conditioned by age, confirming that older students present better skills. Chen et al. [42] analyze differences in the stimulation of CT with a physical robot and a virtual one, concluding that the physical one promotes a greater emotional connection, which facilitates learning. On the other hand, Timur et al. [47] highlight students' ease in humanizing robots.

In terms of the robots used, the most common is Bee-Bot (https://clubbeebot.cl/), a programmable robot designed with a friendly appearance in the shape of a bee. It is easy to use and allows children to program simple sequences of commands (forward,

Tab. 6: Acquisition and development of computational language through block-based programming.

Authorship (date), nationality, stage	Research design		
	Robot objective	Método procedimiento e instrumento (N)	*Robot* objective
Caitité et al. [48], Brazil, primary education	*Line follower robot*: Activate programming skills.	Quantitative: Identification of programming patterns and observation of skills	The use of this robot activates programming skills.
Fanchamps et al. [29], Netherlands, primary education	*Lego NXT* and *Mindstorms*: Stimulate algorithmic thinking (10–12 years).	Experimental quantitative: Algorithmic problem-solving and questionnaire response (N = 62)	The time spent using the robot following its guidelines contributed to competence improvement.
Hsiao et al. [49], Taiwan, primary education	*Arduino* and *Scratch*: Stimulate computational skills to create "Robot Crab" (11–12 years).	Quasi-experimental mixed-method: Programming activities and evaluation with pretest and posttest (N = 70)	Improved computational skills, motivation, and performance.
Martínez et al. [50], Mexico, primary education	*Lego Mindstorms EV3*. Understand if age (6–11 years) determines the proper use of a robot.	Mixed: Creation of a wizard of Oz-based maze: A questionnaire and semistructured interview were used (N = 174)	Older participants develop their social and computational skills better.
Miranda [51], Portugal, early childhood education	*KIBO*: Teach programming skills through reading a classic tale (4–6 years).	Quantitative: Observation of their ability to program the robot and navigate and interact through the story's scenarios	Acquired programming, literature, art, and music skills.
Noh and Lee [52], South Korea, primary education	*Entry* and *Hamster Robot*: Analyze the impact of a programming course to acquire computational skills (10–12 years).	Quantitative: Experimental research (pretest/posttest) after a programming course for 11 weeks (N = 155)	Significant improvement in computational skills.
Sáez et al. [53], Spain, primary education	*mBot*: Analyze the possibilities of a block programmer (11–12 years).	Mixed: Comparison of a group performing mathematical activities with a robot and a control group (N = 93)	Those who used the robot significantly improved their results.
Sullivan and Bers [54], USA, primary education	*KIBO*: Analyze attitudes toward robotics (5–7 years).	Quantitative: Robot implementation for 7 weeks (pretest/posttest) (N = 105)	Increased interest in technology among girls.

Tab. 6 (continued)

| Authorship (date), nationality, stage | Research design | | |
	Robot objective	Método procedimiento e instrumento (N)	*Robot* objective
Yesharim and Ben-Ari [55], Israel, primary education	*Thymio*: Stimulate computational thinking (7–8 years).	Qualitative: Observation of sessions and record on four questionnaires (N = 118)	They learned basic concepts but found it difficult to create and execute their own programs.
Zviel-Girshin et al. [56], Israel, early childhood education	*LEGO® Education WeDo*: Analyze children's perception of robotics and computational language (5–7 years).	Mixed: Semistructured interview: combines a survey (dichotomous items) and open-ended questions (N = 114)	Computational thinking-related skills increase. The gender gap in STEAM areas is reduced.

Source: Own elaboration.

backward, turn left, turn right) to navigate maps and solve problems. Others introduce the STEAM methodology with KIBO (https://kinderlabrobotics.com/kibo/), which is programmed by assembling physical blocks. Similarly, LEGO Education WeDo (https://www.robotix.es/es/lego-wedo) is an educational robotics kit that uses an intuitive visual programming environment to build and program simple robots.

Robotics kits like LEGO Mindstorms EV3 (https://cutt.ly/Nera1wES) and LEGO NXT (https://cutt.ly/rera0csc) are used to build and program robots from a wide range of parts, sensors, and motors, combined with intuitive programming software. They also use mBot (https://www.robotix.es/es/mbot) for block programming and to introduce electronics. Programming platforms like Arduino (https://www.arduino.cc/), open-source based on hardware and software, and Scratch (https://scratch.mit.edu/), visual programming, allow children to create stories, games, and animations, and are used in STEAM projects. They promote logical thinking and introduce programming. There are extensions and projects that link Scratch with robotic hardware, promoting interaction between on-screen programming and the physical world. Some experiment with the Line Follower Robot vehicle (https://cutt.ly/Eera2BCA), programmed to follow a visible line on the ground, and implemented in STEAM projects.

Other research uses humanoid robots like Nao (https://www.aldebaran.com/en/nao), programmed to perform various tasks, from conversation to executing complex movements. They also use the Hamster Robot (https://cogniteam.com/hamster) to introduce basic coding concepts through games and challenges. Additionally, Thymio (https://www.thymio.org/), which incorporates a variety of sensors and preprogrammed modes, help children discover robotics concepts and program visual environments.

4.3 Overall assessment

After analyzing the 44 studies that met the established criteria, it is observed that nearly half adopt quantitative designs (47.7%), some studies are exploratory and descriptive [39], others are quasi-experimental [25, 49, 53] and some are experimental [19, 29, 52]. On the other hand, there are studies with qualitative designs (31.8%), supported by case studies [32, 33] or ethnographic studies [23]. To a lesser extent, other research adopts mixed methodologies (20.5%). However, most do not explicitly mention the research design followed, although it can be inferred (Fig. 3).

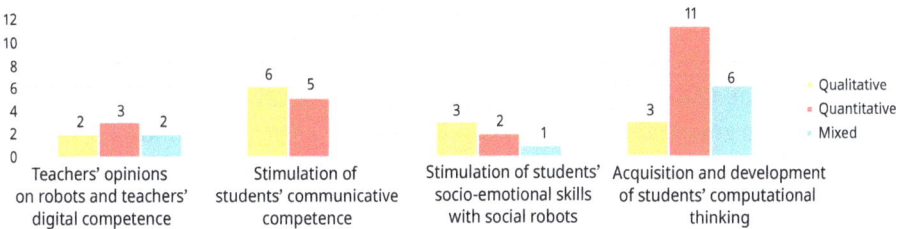

Fig. 3: Distribution of research according to the methodological design.
Source: Authors' own elaboration.

The instruments used to gather information are conditioned by the type of design. Thus, quantitative research primarily employs questionnaires and assessment tests. Questionnaires are used to collect students' opinions on the use of educational robots [29] and teachers' opinions [16]. Assessment tests are employed to analyze to what extent robots promote the enhancement of specific skills and competencies [38]. Some studies opt to compare results between an experimental group using the robot and a control group following a conventional methodology [19, 46]. Others compare scores obtained by the same group before implementing the educational robot (pretest) and after the intervention (posttest) [49]. Assessment tests for specific skills may be standardized [44] or designed ad hoc [49].

On the other hand, qualitative research relies on open or semistructured interviews to gather students' perceptions regarding the pedagogical use of educational robots [43] or teachers' perceptions [20]. Depending on the subjects' age, other techniques such as focus groups or discussion groups are employed. However, observation predominates due to the limitations that early childhood education children may have in responding in writing to certain questionnaire formats. Observation is mainly used to record student interactions with the robot as well as the responses produced [23], the skills and/or competencies acquired and/or developed [27], and/or the motivation generated by the task mediated by the robot [40] (Fig. 4).

Regarding the sample sizes, most studies have small or medium-sized samples, with large samples being exceptional. It is noteworthy that 10 studies do not specify the

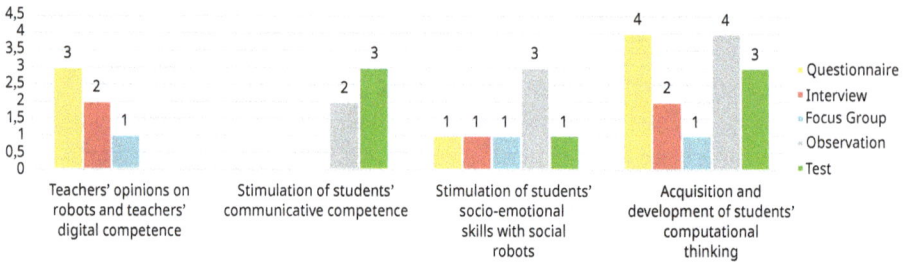

Fig. 4: Distribution of research according to the techniques and instruments used.
Source: Authors' own elaboration.

sample size. There are studies with large samples such as Chootongchai et al. [20] with 510 subjects, and Chen et al. [42] with 767 participants. Meanwhile, others experiment with very small groups, becoming case studies, such as Michaelis and Mutlu [33] (Fig. 5).

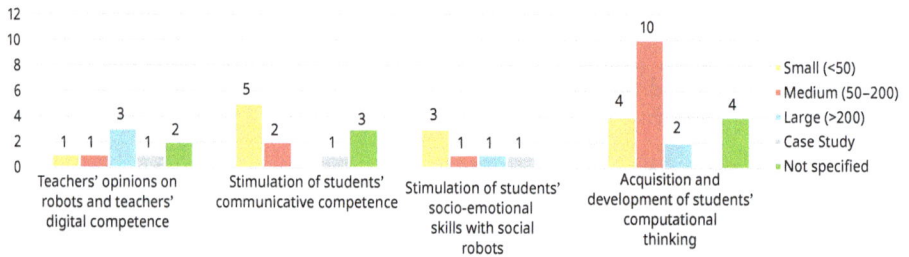

Fig. 5: Distribution of research according to the sample sizes.
Source: Authors' own elaboration.

Regarding the geographical origin of the research groups leading the investigations, it is observed that 20.5% come from the USA, 13.6% from Netherlands, 11.4% from Spain, and 9.1% from Sweden, with smaller percentages distributed among different European countries reaching 20.5%. The Asian representation is 20.4%, and American countries are represented by only 4.6%.

5 Discussion

This review of research has identified the various purposes with which educational robotics is approached in the school context, highlighting the educational opportunities it offers. The main objective of using robots is to develop computational language skills in students, in addition to stimulating other skills such as communication, emotional, narrative, and digital. Most studies describe interventions where the robot takes on a special role, yielding satisfactory results. In some cases, it has contributed to fostering reading habits in children, stimulating language development, increasing vocabulary, etc., undoubtedly linked to the motivation that this novel instrument brings to the classroom context. On the other hand, studies regarding the impact of robots on improving students' socio-emotional skills note the limitations of robots in recognizing emotions. However, they promote autonomy development in certain tasks and motivation toward learning. Additionally, some of the research highlights the opportunity that educational robotics offers to spark young people's interest in programming as well as to increase their computational skills, improve overall academic performance, and in particular, develop their visuospatial memory, self-control, problem-solving skills, teamwork, and interest in technology.

The samples of subjects used are small, and in most cases, they are aimed at testing prototypes, as Poolsawad et al. [16] do. Several studies integrate robots as guides to support STEAM learning in areas such as mathematics or natural sciences [56]. Others analyze the impact of specific robots in enhancing communication skills, whether in the native or foreign language, focusing on increasing vocabulary and promoting reading habits. Some opt for the implementation of social robots to stimulate socio-emotional skills and examine students' responses and perceptions when interacting with the robot. Furthermore, studies focusing on teachers' opinions emphasize their pedagogical utility for promoting various skills, although they acknowledge their lack of digital competence.

Robots used in early childhood education interventions are simpler and more intuitive, with direct physical controls and very basic programming interfaces. They are resistant to falls and impacts, have large and durable parts, and are safe for children of these ages, minimizing the risk of choking. Additionally, they are used to introduce basic concepts of logic, sequencing, cause and effect, and sensory exploration. They do not require advanced reading or math skills, adapting to children's capacities. They usually offer immediate visual or auditory feedback to maintain their engagement with the proposed tasks. On the other hand, robots implemented in Primary Education generally introduce more advanced concepts of programming and electronics, allowing for greater customization. They incorporate more sophisticated programming languages such as Arduino or Scratch.

The seven studies focused on teachers' opinions agree in highlighting the pedagogical usefulness of educational robotics but demand more training to incorporate it into their classrooms, aware of its limitations both methodologically and digitally. Additionally, some express certain ethical concerns about incorporating autonomous robots in the classroom, considering that there are educational tasks exclusive to teachers that cannot be replaced by machines, even if they are intelligent, in the same sense as expressed by Singh et al. [57].

6 Conclusions

There are few studies focused on analyzing teachers' digital competence to incorporate robots into the classroom, and consequently, guidelines are needed for their didactic exploitation, as stated in the Horizon Report [1]. Thus, based on the results of the analyzed research on educational robotics in the educational context, the key to achieving optimal results lies in teacher training. The incorporation of robots requires the implementation of a systematized didactic methodology aimed at achieving the initial educational objectives, designing creative activities that allow for reflective interaction of children with the robot, as well as developing instruments to measure the increase in various skills and/or competencies to be developed, etc.

There are also a few studies that analyze students' perceptions of their interaction with robots. From the results of the systematic review, there is an inference of the need for research to describe the evaluation instruments used, as this would facilitate their extrapolation to other contexts. It would be desirable for them to use larger samples, and for a network of researchers on this topic to be created to identify intervention models and facilitate access to prototypes designed to be extrapolated to other interventions, thus enabling comparison of results. In any case, this review is conditioned by the search period, with the likelihood that more recent and updated studies have been published. However, it can serve as a framework for future research.

Finally, as pointed out by Gómez-León [58], educators must question the role that these artifacts should play, as we are faced with the great challenge of designing robots capable of generating rewarding social interactions supported by social schemes, behaviors, and emotions that promote the socio-emotional education of the children with whom they interact, to promote enriching learning situations that contribute to enhancing their competency development, as well as critical thinking.

References

[1] Pelletier K, McCormack M, Reeves J, Robert J, Arbino N, Al-Freih M, Dickson C, Guevara C, Koster L, Sánchez M, Skallerup L, Stine J. Horizon report, teaching and learning edition. Boulder, CO: EDUCAUSE; 2022.

[2] UNESCO. Framework of competencies for teachers in ICT. UNESCO; 2019. https://unesdoc.unesco.org/ark:/48223/pf0000371024

[3] Wing JM. Computational thinking. Communications of the ACM, 2006, 49(3), 33–35. https://doi.org/10.1145/1118178.1118215

[4] Vuorikari R, Kluzer S, Punie Y. DigComp 2.2: The digital competence framework for citizens-with new examples of knowledge, skills and attitudes. Publications Office of the European Union; 2022. http://doi.org/10.2760/490274

[5] Canbeldek M, Isikoglu N. Exploring the effects of "productive children: Coding and robotics education program" in early childhood education. Education and Information Technologies, 2023, 28(3), 3359–3379. https://doi.org/10.1007/s10639-022-11315-x

[6] Fridberg M, Redfors A, Greca IM, Terceño EM. Spanish and Swedish teachers' perspective of teaching STEM and robotics in preschool–results from the botSTEM project. International Journal of Technology and Design Education, 2023, 33(1), 1–21. https://doi.org/10.1007/s10798-021-09717-y

[7] Darmawansah D, Hwang GJ, Chen MR, Liang JC. Trends and research foci of robotics-based STEM education: A systematic review from diverse angles based on the technology-based learning model. International Journal of STEM Education, 2023, 10(1), 1–24. https://doi.org/10.1186/s40594-023-00400-3

[8] Broza O, Biberman L, Chamo N. "Start from scratch": Integrating computational thinking skills in teacher education program. Thinking Skills and Creativity, 2023, 48, 101285. https://doi.org/10.1016/j.tsc.2023.101285

[9] Su J, Yang W. A systematic review of integrating computational thinking in early childhood education. Computers and Education Open, 2023, 4, 100122. https://doi.org/10.1016/j.caeo.2023.100122

[10] Bers MU, Blake-West J, Kapoor MG, Levinson T, Relkin E, Unahalekhaka A, Yang Z. Coding as another language: Research-based curriculum for early childhood computer science. Early Childhood Research Quarterly, 2023, 64, 394–404. https://doi.org/10.1016/j.ecresq.2023.05.002

[11] Cecchi LA, Rodríguez JP, Dahl V. Logic Programming at Elementary School: Why, What and How Should We Teach Logic Programming to Children?. In: Warren DS, Dahl V, Eiter T, Hermenegildo MV, Kowalski R, Rossi F, eds. Prolog: The next 50 years. Cham: Springer; 2023. pp. 131–143. https://doi.org/10.1007/978-3-031-35254-6_11

[12] Ortega B, Lázaro-Alcalde M. Teachers' perception about the difficulty and use of programming and robotics in the classroom. Interactive Learning Environments, 2023, 31(10), 7074–7085. https://doi.org/10.1080/10494820.2022.2061007

[13] Samara V, Kotsis KT. Primary school teachers' perceptions of using STEM in the classroom attitudes, obstacles, and suggestions: A literature review. Contemporary Mathematics and Science Education, 2023, 4(2), ep23018. https://doi.org/10.30935/conmaths/13298

[14] Shahmoradi S, Kothiyal A, Bruno B, Dillenbourg P. Evaluation of teachers' orchestration tools usage in robotic classrooms. Education and Information Technologies, 2023, 1–38. https://doi.org/10.1007/s10639-023-11909-z

[15] Serholt S, Barendregt W, Vasalou A, Alves-Oliveira P, Jones A, Petisca S, Paiva A. The case of classroom robots: Teachers' deliberations on the ethical tensions. Ai Society, 2017, 32, 613–631. https://doi.org/10.1007/s00146-016-0667-2

[16] Poolsawad K, Songkram N, Piromsopa K, Songkram N. Teachers' perception for integrating educational robots and use as teaching assistants in thai primary sils. Emerging Science Journal, 2020, 4, 127–140. http://dx.doi.org/10.28991/esj-2021-SP1-09

[17] Sánchez-Rivas E, Ruiz-Roso C, Ruiz-Palmero J. Teacher digital competence: Analysis in block programming applied to educational robotics. Sustainability, 2024, 16, 275. https://doi.org/10.3390/su16010275

[18] Gavrilas L, Kotsis KT, Papanikolaou MS. Exploration of the prospective utilization of educational robotics by preschool and primary education teachers. Pedagogical Research, 2024, 9(1), https://doi.org/10.29333/pr/14049

[19] Janicki N, Tenberge C. Technology education in elementary school using the example of learning robots – Development and evaluation of an in-service teacher training concept. Australasian Journal of Technology Education, 2023, 9. https://doi.org/10.15663/ajte.v9.i0.103

[20] Chootongchai S, Songkram N, Piromsopa K. Dimensions of robotic education quality: Teachers' perspectives as teaching assistants in Thai elementary schools. Education and Information Technologies, 2021, 26, 1387–1407. https://doi.org/10.1007/s10639-019-10041-1

[21] Georgieva-Tsaneva G, Andreeva A, Tsvetkova P, Lekova A, Simonska M, Stancheva-Popkostadinova V, Dimitrov G, Rasheva-Yordanova K, Kostadinova I. Exploring the potential of social robots for

speech and language therapy: A review and analysis of interactive scenarios. Machines, 2023, 11(7), 693. https://doi.org/10.3390/machines11070693

[22] Damit SN, Yunus S. Effects of Incorporating Tutor Robot and Game-based Learning for Teaching Mathematics to Primary School Students. In: Ali MY, Karri RR, Shams S, Rosli R, Hj EK, Rahman A, Singh R, eds. 8th Brunei international conference on engineering and technology 2021: Conference proceedings. Vol. 2643, No. 1 Bandar Seri Begawan, Brunei: AIP Publishing; 2023. https://doi.org/10.1063/5.0129978

[23] Ioannou M, Bratitsis T. Utilizing Sphero for a Speed Related STEM Activity in Kindergarten. In: Kolozea E, Panagiotakopoulus C, Skordoulis C, eds. Innovating STEM education: Increased engagement and best practices. Champaign, IL: TheLearner; 2022. pp. 109–118.

[24] Robben D, Fukuda E, De Haas M. The Effect of Gender on Perceived Anthropomorphism and Intentional Acceptance of a Storytelling Robot. In: Castellano G, Riek L, Cakmak M, Leite I, eds. HRI'23 companion of the 2023 ACM/IEEE international conference on human-robot interaction. New York, NY: Association for Computing Machinery; 2023. pp. 495–499. https://doi.org/10.1145/3568294.3580134

[25] Sung J, Lee JY, Chun HY. Short-term effects of a classroom-based STEAM program using robotic kits on children in South Korea. International Journal of STEM Education, 2023, 10(26), https://doi.org/10.1186/s40594-023-00417-8

[26] Samuelsson R. A shape of play to come: Exploring children's play and imaginaries with robots and AI. Computers and Education: Artificial Intelligence, 2023, 5(3), https://doi.org/10.1016/j.caeai.2023.100173

[27] Chatzopoulos A, Kalogiannakis M, Papoutsidakis M, Psycharis S, Papachristos D. Measuring the Impact on Student's Computational Thinking Skills through STEM and Educational Robotics Project Implementation. In: Kalogiannakis M, Papadakis S, eds. Handbook of research on tools for teaching computational thinking in P-12 education. New York, NY: IGI Global; 2020. pp. 238–288. http://doi.org/10.4018/978-1-7998-4576-8.ch010

[28] Jones A, Castellano G. Adaptive robotic tutors that support self-regulated learning: A longer-term investigation with primary school children. International Journal of Social Robotics, 2018, 10, 357–370. https://doi.org/10.1007/s12369-017-0458-z

[29] Fanchamps NL, Slangen L, Hennissen P, Specht M. The influence of SRA programming on algorithmic thinking and self-efficacy using Lego robotics in two types of instruction. International Journal of Technology and Design Education, 2021, 31, 203–222. https://doi.org/10.1007/s10798-019-09559-9

[30] Moher D, Shamseer L, Clarke M, Ghersi D, Liberati A, Petticrew M, Shekelle P, Stewart LA. Preferred reporting items for systematic review and meta-analysis protocols (PRISMA-P) 2015 statement. Systematic Reviews, 2015, 4(1), https://doi.org/10.1186/2046-4053-4-1

[31] Alexander PA. Methodological guidance paper: The art and science of quality systematic reviews. Review of Educational Research, 2020, 90(1), 6–23. http://doi.org/10.3102/0034654319854352

[32] Cho E, Lee K, Cherniak S, Jung SE. Heterogeneous associations of second-graders' learning in robotics class. Technology, Knowledge and Learning, 2017, 22, 465–483. https://doi.org/10.1007/s10758-017-9322-3

[33] Michaelis JE, Mutlu B. Someone to Read With: Design of and Experiences with an In-home Learning Companion Robot for Reading. In: Mark G, Fussell S, Lampe C, Schraefel MC, Hourcade JP, Appert A, Wigdor D, eds. Proceedings of the 2017 CHI conference on human factors in computing systems 2017. New York, NY: Association for Computing Machinery; 2017. pp. 301–312. https://doi.org/10.1145/3025453.3025499

[34] Moreno V, Rodriguez-Muñoz FJ. Design and piloting of a proposal for intervention with educational robotics for the development of lexical relationships in early childhood education. Smart Learning Environments, 2023, 10(6), https://doi.org/10.1186/s40561-023-00226-0

[35] Westlund JMK, Dickens L, Jeong S, Harris PL, DeSteno D, Breazeal CL. Children use non-verbal cues to learn new words from robots as well as people. International Journal of Child-Computer Interaction, 2017, 13, 1–9. https://doi.org/10.1016/j.ijcci.2017.04.001

[36] Westlund K, Jacqueline M, Jeong S, Park HW, Ronfard S, Adhikari A, Breazeal CL. Flat vs. expressive storytelling: Young children's learning and retention of a social robot's narrative. Frontiers in Human Neuroscience, 2017, 11(295), http://doi.org/10.3389/fnhum.2017.00295

[37] Alemi M, Meghdari A, Haer NS. Young EFL Learners' Attitude Towards RALL: An Observational Study Focusing on Motivation, Anxiety, and Interaction. In: Kheddar A, Yoshida E, Ge SS, Suzuki K, Cabibihan JJ, Eyssel F, He H, eds. Social robotics: 9th international conference, ICSR 2017. Tsukuba, Japan, November 22–24, 2017, Proceedings Tsukuba: Springer International Publishing; 2017. pp. 252–261. https://doi.org/10.1007/978-3-319-70022-9_25

[38] De Wit J, Schodde T, Willemsen B, Bergmann K, De Haas M, Kopp S, Vogt P. The Effect of a Robot's Gestures and Adaptive Tutoring on Children's Acquisition of Second Language Vocabularies. In: Kanda T, Šabanović S, Hoffman G, Tapus A, eds. Proceedings of the 2018 ACM/IEEE international conference on human-robot interaction 2018. New York, NY: Association for Computing Machinery; 2018. pp. 50–58. https://doi.org/10.1145/3171221.3171277

[39] Haas MD, Baxter P, de Jong C, Krahmer E, Vogt P. Exploring Different Types of Feedback in Preschooler and Robot Interaction. In: Mutlu B, Tscheligi M, Weiss A, Young JE, eds. Proceedings of the companion of the 2017 acm/ieee international conference on human-robot interaction. New York, NY: Association for Computing Machinery; 2017. pp. 127–128. https://doi.org/10.1145/3029798.3038433

[40] Rintjema E, Van Den Berghe R, Kessels A, De Wit J, Vogt P. A Robot Teaching Young Children A Second Language: The Effect of Multiple Interactions on Engagement and Performance. In: Kanda T, Šabanović S, Hoffman G, Tapus A, eds. HRI'18 companion of the 2018 ACM/IEEE international conference on human-robot interaction 2018. New York, NY: Association for Computing Machinery; 2018. pp. 219–220. https://doi.org/10.1145/3173386.3177059

[41] Caballero YA, Muñoz L, Muñoz AG. Pilot Experience: Play and Program with Bee-bot to Foster Computational Thinking Learning in Young Children. In: International engineering, sciences and technology conference. 2019 7th international engineering, sciences and technology conference (IESTEC): Proceedings. IEEE Computer Society; 2019. pp. 601–606. https://doi.org/10.1109/IESTEC46403.2019.00113

[42] Chen G, Shen J, Barth-Cohen L, Jiang S, Huang X, Eltoukhy M. Assessing elementary students' computational thinking in everyday reasoning and robotics programming. Computers Education, 2017, 109, 162–175. https://doi.org/10.1016/j.compedu.2017.03.001

[43] Ching YH, Yang D, Wang S, Baek Y, Swanson S, Chittoori B. Elementary school student development of STEM attitudes and perceived learning in a STEM integrated robotics curriculum. TechTrends, 2019, 63, 590–601. https://doi.org/10.1007/s11528-019-00388-0

[44] Di Lieto MC, Inguaggiato E, Castro E, Cecchi F, Cioni G, Dell'Omo M, Dario P. Educational robotics intervention on executive functions in preschool children: A pilot study. Computers in Human Behaviour, 2017, 71, 16–23. https://doi.org/10.1016/j.chb.2017.01.018

[45] Kopcha TJ, McGregor J, Shin S, Qian Y, Choi J, Hill R, Choi I. Developing an integrative STEM curriculum for robotics education through educational design research. Journal of Formative Design in Learning, 2017, 1, 31–44. https://doi.org/10.1007/s41686-017-0005-1

[46] Sáez JM, Buceta R, De Lara S. La aplicación de la robótica y programación por bloques en la enseñanza elemental. RIED-Revista Iberoamericana De Educación a Distancia, 2021, 24(1), 95–113. https://doi.org/10.5944/ried.24.1.27649

[47] Timur B, Timur S, Güvenç E, Önder EY. Primary school students' perceptions about robotic coding. Journal of Individual Differences in Education, 2021, 3(1), 20–29. https://doi.org/10.47156/jide.953229

[48] Caitité VGR, dos Santos DMG, Gregório IC, da Silva WB, Mendes VF. Diffusion of Robotics through Line Follower Robots. In: Pessoa J, eds. 2018 Latin American robotic symposium, 2018 Brazilian symposium on robotics (SBR) and 2018 workshop on robotics in education (WRE). Brazil: Institute of Electrical and Electronics Engineers (IEEE); 2018. pp. 604–609. http://doi.org/10.1109/LARS/SBR/WRE.2018.00109

[49] Hsiao HS, Lin YW, Lin KY, Lin CY, Chen JH, Chen JC. Using robot-based practices to develop an activity that incorporated the 6E model to improve elementary school students' learning performances. Interactive Learning Environments, 2022, 30(1), 85–99. https://doi.org/10.1080/10494820.2019.1636090

[50] Martínez J, Pérez H, Espinosa I, Avila H, Rodríguez J. Age-based differences in preferences and affective reactions towards a robot's personality during interaction. Computers in Human Behavior, 2018, 84, 245–257. https://doi.org/10.1016/j.chb.2018.02.039

[51] Miranda MS. Powerful Ideas and the Kibo Robot Curriculum: The Traditional Children's Stories, for the Integration of Programming and Robotics. In: Gómez-Chova L, López-Martínez A, Candel-Torres I, eds. EDULEARN21 proceedings 13th international conference on education and new learning technologies. Palma: IATED Academy; 2021. pp. 3595–3604. https://doi.org/10.21125/edulearn.2021.0755

[52] Noh J, Lee J. Effects of robotics programming on the computational thinking and creativity of elementary school students. Educational Technology Research and Development, 2020, 68, 463–484. https://doi.org/10.1007/s11423-019-09708-w

[53] Sáez JM, Sevillano ML, Vazquez E. The effect of programming on primary school students' mathematical and scientific understanding: Educational use of mBot. Educational Technology Research and Development, 2019, 67, 1405–1425. https://doi.org/10.1007/s11423-019-09648-5

[54] Sullivan A, Bers MU. Investigating the use of robotics to increase girls' interest in engineering during early elementary school. International Journal of Technology and Design Education, 2019, 29, 1033–1051. https://doi.org/10.1007/s10798-018-9483-y

[55] Yesharim MF, Ben-Ari M. Teaching computer science concepts through robotics to elementary school children. International Journal of Computer Science Education in Schools, 2018, 2(3), https://doi.org/10.21585/ijcses.v2i3.30

[56] Zviel-Girshin RZ, Rosenberg N, Kukliansky I. Early childhood robotics: Children's beliefs and objective capabilities to read and write programs. Journal of Research in Childhood Education, 2023. https://doi.org/10.1080/02568543.2023.2259946

[57] Singh DK, Kumar M, Fosch-Villaronga E, Singh D, Shukla J. Ethical considerations from child-robot interactions in under-resourced communities. International Journal of Social Robotics, 2023, 15(12), 2055–2071. https://doi.org/10.1007/s12369-022-00882-1

[58] Gómez-León MI. Robots sociales y crecimiento ético en Educación Infantil. Edutec Revista Electrónica de Tecnología Educativa, 2023, 83, 41–54. https://doi.org/10.21556/edutec.2023.83.2697

[59] Baxter P, Lightbody P, Hanheide M. Robots providing cognitive assistance in shared workspaces. In: Companion of the 2018 ACM/IEEE International Conference on Human-Robot Interaction. 2018; p. 57–58. https://doi.org/10.1145/3173386.3177070

[60] Heljakka K, Lamminen A, Ihamäki P. A model for enhancing emotional literacy through playful learning with a robot dog. In: 2021 International Conference on Electrical, Computer, Communications and Mechatronics Engineering (ICECCME). IEE. 2018; p. 1–7. http://doi.org/10.1109/ICECCME52200.2021.9590996

[61] Serholt S. Breakdowns in children's interactions with a robotic tutor: A longitudinal study. Computers in Human Behavior. 2018; 81: p. 250–264. https://doi.org/10.1016/j.chb.2017.12.030

[62] Serholt S. Interactions with an empathic robot tutor in education: students' perceptions three years later. In: Artificial Intelligence and Inclusive Education: Speculative Futures and Emerging Practices. Springer. 2019; p. 77–99. https://doi.org/10.1007/978-981-13-8161-4_5

[63] Sánchez-Rivas E, Ruiz-Roso Vázquez C, Ruiz-Palmero J. Teacher Digital Competence Analysis in Block Programming Applied to Educational Robotics. Sustainability. 2023; 16(1), p. 275. https://doi.org/10.3390/su16010275

[64] Van den Bergh J, Van Deurzen B, Veuskens T, Ramakers R, & Luyten K. Towards tool-support for robot-assisted product creation in Fab Labs. In: International Conference on Human-Centred Software Engineering. Cham: Springer International Publishing. 2018; p. 2019–230. https://doi.org/10.1007/978-3-030-05909-5_13

[65] Michaelis JE, & Mutlu B. Supporting interest in science learning with a social robot. In: Proceedings of the 18th ACM International Conference on Interaction Design and Children. 2019; p. 71–82. https://doi.org/10.1145/3311927.3323154

Daniela Conti, Carla Cirasa, and Santo F. Di Nuovo

The use of psychological measurements in research on educational robotics

Abstract: In recent decades, there has been an increase in the use of educational robots in classrooms with K-12 students. The objective of employing educational robotics in educational environments is to apply robotics to promote various skills among children and adolescents, including social, communication, cognitive, learning, and other achievement skills. Both K-12 students and teachers have displayed positive attitudes toward educational robotics, highlighting its potential as a valuable tool in subjects like science, programming, and mathematics. Research on the effectiveness and attitudes toward educational robots has also demonstrated promising results. However, the majority of the conducted research lacks quantitative evaluations using standardized testing. That is, while psychometrics is commonly employed in research within various educational contexts, the use of psychometrics in educational robotics research is scarce. Because many robots developed for educational contexts aim to achieve learning objectives similar to those encountered in human interactions, it becomes advantageous to use psychological measurement tools in the evaluation process, specifically when assessing psychological outcome objectives. This inclusion allows for a better understanding of the educational robot's effectiveness on psychological outcomes, contributing to a more informed and insightful assessment of their impact on educational settings. This chapter will provide an overview of some psychological measurement tools used in educational robotics research. Identifying some standardized psychological tests that are used in the research of human-robot interactions can offer insights into the potential suitability of such tests for further assessing educational robotics.

Keywords: Educational robots, human-robot interactions, psychometrics

1 Introduction

An educational social robot can be defined as a robot that, through social interaction, contributes to providing a learning experience to students. This differentiates them from the robots used for the STEM method, which are tools for scientific, technological, engi-

Corresponding author: **Santo F. Di Nuovo**, Department of Educational Sciences, University of Catania, Catania, Italy, e-mail: s.dinuovo@unict.it

Daniela Conti, Department of Humanities, University of Catania, Catania, Italy,
e-mail: daniela.conti@unict.it

Carla Cirasa, Department of Humanities, University of Catania, Catania, Italy,
e-mail: carla.cirasa@phd.unict.it

https://doi.org/10.1515/9783111352695-005

neering, and mathematics education that are not able to enter a relationship with the child. Social robots have been designed and constructed to engage with humans in different contexts of their lives. Their primary purpose is not task-oriented but rather to convey stimuli beneficial to social interactions. This interaction is facilitated by the robots' characteristics, which closely resemble those of humans. In fact, embeddedness and anthropomorphism represent the main advantages of using these robots [1]. Due to robots' human-like characteristics, such as their eyes and mouths, social robots can play an essential role in fostering natural interactions between humans and robots [2]. Moreover, they prove to be more conducive to stimulating dialog and language development compared to devices lacking these anthropomorphic features [3]. Various types of social robots are currently available with different degrees of freedom and of various sizes. Some have been designed for specific purposes, while others are programmable and customizable to your liking. As suggested in a review by Youssef et al. [4], their application in the field of childhood ranges from medicine [5], to care and assistance [6], to telepresence [7], to accompaniment [8], and education [9]. In the field of education, social robots have been applied to perform various roles, both as peers and as tutors [10]. Furthermore, social robots can also play the role of human teacher assistants, autonomous tutors, partners with whom to carry out a task, or opponents in a game [2]. In a study by Golonka et al. [11], it has in fact been demonstrated that the robot tutor is perceived as less intimidating than peers and the teacher. In the field of teaching, their major application has been storytelling [12, 13] during this activity, they perform better than a tablet or other sound support as they are able to transmit social behaviors and use facial expressions and gestures [14]. These skills are also useful in learning foreign languages [15]. In fact, when learning a new language, interaction and repeated practice are essential to improve. The same applies to reading ability [16, 17]. In recent years, social robots are being used in the field of education also for transversal knowledge such as food education [18] or correct waste disposal [8]. An emerging line of studies linked to the use of robotics in teaching is that linked to emotions. In fact, recent studies have questioned the ability of these robots to transmit emotions to the child with whom they interact and what possible individual differences can be found in the ability of children to correctly read the emotional stages of the robot. This ability is of primary importance in the perspective of programming increasingly performative robots capable of building profitable relationships. Konjin et al. [19] showed how, when interacting with humanoids, people can feel an emotional bond between themselves and the robot. This bond can therefore become a driving force for greater involvement and consequently learning.

The remaining parts of the chapter are organized as follows: Section 2 reviews the scientific literature that contextualizes this work, focusing on measurements in human-robot interaction; Section 3 presents the psychological measurements, that is, surveys, self-reports, and standardized questionnaires; and Section 4 discusses the conclusions and directions for future research.

2 Measurements in human-robot interactions

Social robots can interact with children in various ways, both on an individual level [15, 20] or in group contexts [13, 21]. In the recent decade, various studies have tested the benefits of using robots in educational settings. For example, some studies have evaluated the advantages of using robotic systems compared to other types of technological tools such as tablets. In a study by Yang [22], two samples of preschool children were compared. One group completed a learning program with the robot, while the other group completed the same program using a tablet. The group who completed the program with the robot had an improvement in emotional regulation and social skills. Similar findings were demonstrated by Konjin et al. [23] in 2022, where higher levels of performance, engagement, and enjoyment were noticed among the children who interacted with a robot. The interactions between students and robots are also influenced by how the robot responds to students. In a study by de Haas [24], 5–6 year-olds took part in three learning sessions, each with different levels and degrees of feedback from the robot. The children in the group that received the most feedback and guidance from the robot were more engaged with the robot. Furthermore, it has been shown that the use of robotic gestures during interactions with humans, compared to a static robot, increases both involvement and learning [25]. It is well-known that nonverbal behavior, such as changing the direction of gaze, facial expressions, and gestures, are important aspects of communication between people. That robots are able to imitate the same characteristics is also of value in order to facilitate an interaction that appears as natural and humanlike as possible [26]. This might contribute to more natural relationships between humans and robots [27]. Furthermore, for the interaction between humans and robots to be natural, it is important to know more about whether humans consider robots as their own beings or just a technological tool without emotions and a personality. In a study by Dio et al. [28], children's ability to attribute mental states to a robot was examined. The results indicated that attachment and Theory of Mind skills contributed to a different extent, depending on the age of development, to the child's internal representation of a robot's mental state. Furthermore, robots can offer opportunities to create an engaging and nonjudgmental educational setting, which can be beneficial to reduce anxiety levels related to school performance [29]. Additionally, social robots might be able to serve as assistants to help students with challenges they face in their education. For example, Cagiltay et al. [30] investigated children's preferences when codesigning a social robot that would be able to help them with their homework. The children communicated that they preferred that the robot supported their emotional state and helped with motivation to complete tasks. The same answers were given by the children's parents, who evaluated the robot's ability to express emotions and the robot's ability to inspire the students' motivation to study to be of significance.

3 Psychological measurements

Using reliable tools that ensure replicability is important to develop credible knowledge within human-robot interactions. As early as 1989, the scarcity of valid and reliable measuring instruments for assessing the usability and acceptance of computers was acknowledged [31]. In recent years, it has been highlighted that much of the research on interactions between humans and robots lacks the use of valid questionnaires and tools [32]. In the same way as when conducting research on humans, the choice of measurements in the research on the interaction between robots and humans depends on the outcome measures to be investigated and the purpose of the investigation. On the other hand, it is of value that measurements of different outcome variables are conducted using instruments that have been tested for validity and reliability. This might also make it easier to replicate the conducted work, which is useful for the field of research.

3.1 Surveys and self-report

Both in cross-sectional studies and in experimental studies conducted on human-human interactions, questionnaires filled out through self-report have been frequently used. Self-reporting enables a rapid collection of responses, which can be recorded in various formats, for example, Likert scales, allowing for the quantification of participants' responses. As Kormos and Gifford [33] state: "Self-reports are the preferred method of data collection for the majority of researchers, owing to their low cost, relative ease of use, and flexibility." Even though self-report is mostly used in research on human-human interactions, the method has also been used in research on robotics. For example, in a study by Akechi et al. [34], participants were asked to assess whether a robot was capable of exercising self-control, feeling fear, remembering, experiencing hunger, and making plans. The participants were asked to rate their answers on a six-point scale from 0 (not at all) to 5 (very much). Similar questions were asked in a study by Ward et al. [35], where the Likert scale was changed from 0–5 to 1–7, with anchors ranging from 1 (extremely wrong) to 7 (extremely right). In a study by Sobel and Sims [36], the lifelikeness of a triangle, presented as a robot, was measured with the perceived humanness index [37], as referred to [38], which includes six different semantic differential scales (e.g., artificial ←→ natural; human-made ←→ humanlike, and indefinite lifespan ←→ mortal).

3.2 Standardized questionnaires

As mentioned earlier, it is important that research on robots is conducted in a way that makes it feasible to replicate. One way to achieve this is by using standardized questionnaires or measurements in the research. Standardized tests are widely used in traditional educational and psychological research. However, standardized measuring instruments and tests have also been used in assessments of HRI. For example, the NEPSY-II test has been employed in robot research. The instrument was used in a study by Brooks et al. [39], and it was revealed that children who interacted with a humanoid robot achieved higher scores on the test compared to those who interacted with a non-humanoid robot. The NEPSY-II is a comprehensive neuropsychological assessment tool designed for children aged 3–12, consisting of 32 subtests assessing various neurocognitive abilities (e.g., social perception, memory, and learning; see also [40]). Other instruments evaluate factors such as empathy, mentalizing, or credibility. Further, the chapter will provide an overview of some well-known psychological tests and experiments that have been used in research on HRI.

3.2.1 False belief tests

The false belief test involves one individual and an object. When the individual leaves the room, the object is moved, and the participants in the study are asked to predict where the individual will look for the object they left behind. A 5-year-old child will recognize that although they have witnessed the object move, the individual has not. Thus, they will give an answer that aligns with where the individual left the object earlier [41]. Zhang et al. [42] used this procedure in their study with children as participants and where the individual was replaced with a robot. In the study, the robot held a ball and stated: "This is my ball. I will put it in this box and come back later to look for it." The participants were asked to move the ball when the robot was not present. After this experimental procedure, the participants were asked: (1) *Where is the ball now?* (2) *Before the NAO robot left, where was the ball?*, and finally, (3) *When the robot comes back, where will he look for the ball first?* In addition to the traditional false belief test, explicit false-belief tests, knowledge-ignorance assessments, contents false beliefs, and hidden emotions (see also [43]) have been used in research on HRI [44].

3.2.2 Perspective taking

The ability to understand how others feel in specified situations is influenced by different factors. The ability to recognize others' feelings is crucial for the development of empathy, helping others, and building trust [45]. A common method to assess an individual's perspective-taking ability is to ask them to imagine how another person will feel and then rate their experience of distress or empathy toward that person [46]. Perspective-taking is also vital in research on robots, as the goal is often to make robots appear as human as possible. In a study where an android robot was used to facilitate job interviews with people with autism, perspective-taking was measured with the question, "Do you think your perspective is different from most people's?"

The test subjects rated on a 5-point Likert scale the extent to which they understood the interviewer's point of view [51]. In a study by Scassellati et al. [47], children engaged with a social robot daily for 30 min alongside a caregiver. The activities focused, among others, on perspective-taking, introducing the children to two games, Rocket and House (see more [47]). Related to perspective-taking in research on robots is also the study conducted by Sobel and Sims [36], where participants saw triangles labeled as "humans" or "robots." The results showed that participants attributed more mental qualities to the triangles referred to as humans.

4 Conclusions

Social robots are designed to interact with people in different contexts. Using valid tools when evaluating human-robot interactions will allow future research to delve further into aspects of human-robot interaction that have not yet been investigated. Despite considerable progress, we still have scarce knowledge of how we can best evaluate the interaction between robots and humans, especially when examining K-12 education students. Using valid measurement tools can help to better understand how to design educational interventions with robots that are in line with the needs of students. Simultaneously, the use of robots in educational contexts is still limited by technical and logistical challenges as well as ethical questions. Although the benefits of using robots are widely demonstrated in research on children, adults are still cautious in considering them suitable to use in teaching contexts with children. Indeed, robots currently offer sectorial knowledge, but it is evident that they are not ready to replace a human teacher. We must, to a greater extent, be able to evaluate how many aspects of the interaction between human beings are lost when replacing one human with a robot.

On the one hand, the use of robots in educational environments can offer an important advantage, namely that the robots and their purpose can be adapted based on individual children's needs. However, it is important to mention that several aspects of the relationship between human beings could be lost when using robots. In fact, we are still far from being able to say that the interactions between humans and humans and humans and robots are equivalent. However, it is plausible to assume that, especially in school contexts, relationships between human beings could be mediated by robots. To better investigate the effects and impact robots can have on educational settings with students, it is important that this interaction is investigated. To investigate this, one should to a greater extent put into use valid measurement tools that allow us to evaluate the psychological, affective, and cognitive factors and how these factors influence the relationship between humans and robots. Finally, since these are educational contexts, we need to consider that these factors also need to be considered in developmental contexts, as many of the subjects are in either childhood or adolescence. Specifically, as children grow up, they could change their perspective or develop different relational skills, which could therefore change their relationship with the robot itself. Future research should therefore address these questions by developing instruments that are reliable and valid but also practical and easy to use. In the field of educational robotics, questionnaires are still the most used tools precisely because of these characteristics. However, questionnaires may not be a sufficient method to use to collect the complex information discussed above. Future research should therefore suggest and develop sensitive but also flexible and easy-to-use tools that can be used in real interaction contexts in classrooms or during learning sessions. This can allow research to expand knowledge on the interactions between robots and children in educational contexts, but also allow researchers to gain valuable information for computer scientists who implement the behavioral repertoire of robots in order to create their interactions in learning contexts.

References

[1]   van den Berghe R, de Haas M, Oudgenoeg-Paz O, Krahmer E, Verhagen J, Vogt P, et al. A toy or a friend? Children's anthropomorphic beliefs about robots and how these relate to second-language word learning. Journal of Computer Assisted Learning, 2021, 37(2), 396–410.

[2]   Conti D, Cirasa C, Di Nuovo S, Di Nuovo A. "Robot, tell me a tale!": A social robot as tool for teachers in kindergarten. Interaction Studies: Social Behaviour and Communication in Biological and Artificial Systems, 2020, 21(2), 220–242.

[3]   Randall N. A Survey of Robot-Assisted Language Learning (RALL). Journal Human-Robot Interaction, 2019, 9(1), 7.

[4]   Youssef K, Said S, Alkork S, Beyrouthy T. Social Robotics in Education: A Survey on Recent Studies and Applications. International Journal of Emerging Technologies in Learning (Online), 2023, 18(3), 67.

[5]   Uluer P, Kose H, Gumuslu E, Barkana DE. Experience with an Affective Robot Assistant for Children with Hearing Disabilities. International Journal of Social Robotics, 2023, 15(4), 643–660.

[6]   Frennert S, Aminoff H, Östlund B. Technological Frames and Care Robots in Eldercare. International Journal of Social Robotics, 2021, 13(2), 311–325.

[7]   Tuli TB, Terefe TO, Rashid MMU. Telepresence Mobile Robots Design and Control for Social Interaction. International Journal of Social Robotics, 2021, 13(5), 877–886.

[8]   Castellano G, De Carolis B, D'Errico F, Macchiarulo N, Rossano V. PeppeRecycle: Improving Children's Attitude Toward Recycling by Playing with a Social Robot. International Journal of Social Robotics, 2021, 13(1), 97–111.

[9]   Conti D, Di Nuovo S, Buono S, Di Nuovo A. Robots in education and care of children with developmental disabilities: A study on acceptance by experienced and future professionals. International Journal of Social Robotics, 2021, 9, 51–62.

[10]   Conti D, Trubia G, Buono S, Di Nuovo S, Di Nuovo A. Social robots to support practitioners in the education and clinical care of children: The CARER-AID project. Life Span and Disability, 2020, 23(1), 17–30.

[11]   Golonka EM, Bowles AR, Frank VM, Richardson DL, Freynik S.Technologies for foreign language learning: A review of technology types and their effectiveness. Computer Assisted Language Learning, 2014, 27(1), 70–105.

[12] Augello A, Città G, Gentile M, Lieto A. A Storytelling Robot Managing Persuasive and Ethical Stances via ACT-R: An Exploratory Study. International Journal of Social Robotics, 2023, 15(12), 2115–2131.

[13] Conti D, Nuovo AD, Cirasa C, Nuovo SD. A Comparison of Kindergarten Storytelling by Human and Humanoid Robot with Different Social Behavior. In: Proceedings of the Companion of the 2017 ACM/IEEE International Conference on Human-Robot Interaction. Vienna, Austria: Association for Computing Machinery; 2017. pp. 97–98.

[14] Striepe H, Donnermann M, Lein M, Lugrin B. Modeling and Evaluating Emotion, Contextual Head Movement and Voices for a Social Robot Storyteller. International Journal of Social Robotics, 2021, 13(3), 441–457.

[15] Belpaeme T, Vogt P, van den Berghe R, Bergmann K, Göksun T, de Haas M, et al. Guidelines for Designing Social Robots as Second Language Tutors. International Journal of Social Robotics, 2018, 10(3), 325–341.

[16] Liu X, Ma J, Wang Q. A social robot as your reading companion: Exploring the relationships between gaze patterns and knowledge gains. Journal on Multimodal User Interfaces, 2023, 1, 21–41.

[17] Caruana N, Moffat R, Miguel-Blanco A, Cross ES. Perceptions of intelligence & sentience shape children's interactions with robot reading companions. Scientific Reports, 2023, 13(1), 7341.

[18] Balzotti A, Carolis BD, Massaro S, Perla L, Rossano V. Healthy Pepper: Nutritional Education Through Social Robotics and Storytelling. 2023 IEEE International Conference on Advanced Learning Technologies (ICALT), 2023, 121–123, IEEE.

[19] Konijn EA, Hoorn JF. Parasocial interaction and beyond: Media personae and affective bonding. The International Encyclopedia of Media Effects. New York, NY, USA: Wiley-Blackwell; 2017, 1–15.

[20] de Haas M, Vogt P, Krahmer E. When Preschoolers Interact with an Educational Robot. Does Robot Feedback Influence Engagement? Multimodal Technologies and Interaction, 2021, 5(12), 77.

[21] Al Hakim VG, Yang S-H, Liyanawatta M, Wang J-H, Chen G-D. Robots in situated learning classrooms with immediate feedback mechanisms to improve students' learning performance. Computers & Education, 2022, 182, 104483.

[22] Yang W. Coding With Robots or Tablets? Effects of Technology-Enhanced Embodied Learning on Preschoolers' Computational Thinking and Social-Emotional Competence. Journal of Educational Computing Research, 2024, 0(0), 07356331241226459.

[23] Konijn EA, Jansen B, Mondaca Bustos V, Hobbelink VLNF, Preciado Vanegas D. Social Robots for (Second) Language Learning in (Migrant) Primary School Children. International Journal of Social Robotics, 2022, 14(3), 827–843.

[24] de Haas M, Vogt P, Krahmer E. The Effects of Feedback on Children's Engagement and Learning Outcomes in Robot-Assisted Second Language Learning. Frontiers in Robotics and AI, 2020, 7, 101.

[25] de Haas M, Vogt P, van den Berghe R, Leseman P, Oudgenoeg-Paz O, Willemsen B, et al. Engagement in longitudinal child-robot language learning interactions: Disentangling robot and task engagement. International Journal of Child-Computer Interaction. 2022, 33, 100501.

[26] Zinina A, Zaidelman L, Arinkin N, Kotov A. Non-verbal behavior of the robot companion: A contribution to the likeability. Procedia Computer Science, 2020, 169, 800–806.

[27] Urakami J, Sutthithatip S. Building a Collaborative Relationship between Human and Robot through Verbal and Non-Verbal Interaction. In: Companion of the 2021 ACM/IEEE International Conference on Human-Robot Interaction. Boulder, CO, USA: Association for Computing Machinery; 2021. pp. 257–261.

[28] Di Dio C, Manzi F, Peretti G, Cangelosi A, Harris PL, Massaro D, et al. Come i bambini pensano alla mente del robot. Il ruolo dell'attaccamento e della Teoria della Mente nell'attribuzione di stati mentali ad un agente robotico. Sistemi Intelligenti, 2020, 32(1), 41–56.

[29] Hsu T-C, Chang C, Wu L-K, Looi C-K. Effects of a Pair Programming Educational Robot-Based Approach on Students' Interdisciplinary Learning of Computational Thinking and Language Learning. Frontiers in Psychology, 2022, 13, 888215.

[30] Cagiltay B, Mutlu B, Michaelis JE. "My Unconditional Homework Buddy:" Exploring Children's Preferences for a Homework Companion Robot. In: Proceedings of the 22nd Annual ACM Interaction Design and Children Conference. Chicago, IL, USA: Association for Computing Machinery; 2023, pp. 375–387.

[31] Davis FD. Perceived Usefulness, Perceived Ease of Use, and User Acceptance of Information Technology. MIS Quarterly, 1989, 13(3), 319–340.

[32] Conti D, Di Nuovo S. Assessment by telematic means and artificial agents: A new challenge for psychometrics? BPA Applied Psychology Bulletin, 2023, 296(1), 40–52.

[33] Kormos C, Gifford R. The validity of self-report measures of proenvironmental behavior: A meta-analytic review. Journal of Environmental Psychology, 2014, 40, 359–371.

[34] Akechi H, Kikuchi Y, Tojo Y, Hakarino K, Hasegawa T. Mind perception and moral judgment in autism. Autism Research, 2018, 11(9), 1239–1244.

[35] Ward AF, Olsen AS, Wegner DM. The harm-made mind: Observing victimization augments attribution of minds to vegetative patients, robots, and the dead. Psychological Science, 2013, 24(8), 1437–1445.

[36] Sobel BM, Sims VK. Using the Frith-Happé animations to compare attributions of mental qualities in nonhuman agents. Journal of Cognitive Psychology, 2023, 35(3), 330–339.

[37] Ho CC, MacDorman KF. Revisiting the uncanny valley theory: Developing and validating an alternative to the Godspeed indices. Computers in Human Behavior, 2010, 26(6), 1508–1518.

[38] Akechi H, Kikuchi Y, Tojo Y, Hakarino K, Hasegawa T. Mind perception and moral judgment in autism. Autism Research, 2018, 11(9), 1239–1244.

[39] Brooks BL, Sherman EM, Strauss E. NEPSY-II: A developmental neuropsychological assessment. Child Neuropsychology, 2009, 16(1), 80–101.

[40] Vasilopoulos F, Jeffrey H, Wu Y, Dumontheil I. Multi-level meta-analysis of physical activity interventions during childhood: Effects of physical activity on cognition and academic achievement. Educational Psychology Review, 2023, 35(2), 59.

[41] Wellman HM, Cross D, Watson J. Meta-analysis of theory of mind development: The truth about false belief. Child Development, 2001, 72(3), 655–684.

[42] Zhang Y, Song W, Tan Z, Wang Y, Lam CM, Hoi SP, Yi L. Theory of robot mind: False belief attribution to social robots in children with and without autism. Frontiers in Psychology, 2019, 10, 1732.

[43] Wellman HM, Fang F, Liu D, Zhu L, Liu G. Scaling of theory-of-mind understandings in Chinese children. Psychological Science, 2006, 17(12), 1075–1081.

[44] So WC, Cheng CH, Lam WY, Huang Y, Ng KC, Tung HC, Wong W. A robot-based play-drama intervention may improve the joint attention and functional play behaviors of Chinese-speaking preschoolers with autism spectrum disorder: A pilot study. Journal of Autism and Developmental Disorders, 2020, 50,467–481.

[45] Batson CD, Early S, Salvarani G. Perspective taking: Imagining how another feels versus imaging how you would feel. Personality and Social Psychology Bulletin, 1997, 23(7), 751–758.

[46] Kumazaki H, Muramatsu T, Yoshikawa Y, Matsumoto Y, Ishiguro H, Mimura M, Kikuchi M. Role-play-based guidance for job interviews using an android robot for individuals with autism spectrum disorders. Frontiers in Psychiatry, 2019, 10, 239.

[47] Scassellati B, Boccanfuso L, Huang CM, Mademtzi M, Qin M, Salomons N, Shic F. Improving social skills in children with ASD using a long-term, in-home social robot. Science Robotics, 2018, 3(21), eaat7544.

Rita Cersosimo and Valentina Pennazio

Promoting social communication skills in autism spectrum disorder through robotics and virtual worlds

Abstract: The chapter aims to present the plan of a research project that has the objective of testing how the use of social robots can effectively enhance communication and relational skills in children and adolescents with autism spectrum disorder. After a brief introduction on social communication skills in the spectrum, the chapter will outline previous studies demonstrating that social robots can be useful to open a communicative channel in people with autism, acting as mediators within social situations. Key advantages include the predictability and emotional simplicity of robots, which can be programmed to meet the sensory needs of each individual with autism. Moreover, they allow for the implementation of increasingly complex communicative exchanges. To promote the generalization of skills learned through interaction with the robot, the experimental design also includes the use of virtual environments where individuals can practice skills learned in an initial social context, leading ultimately to the application of these skills in a real-world setting.

Keywords: Social robots, autism spectrum disorder, social communication skills

1 Introduction

For several years now, the international literature [1, Kiuppis & Hausstätter, 2014; 70] has shown how the concept of inclusion is broadening its range of research and application. In the school context, this concept has shifted its attention from individualized and personalized learning paths designed for single students to approaches of the same type designed for all students with the aim of enhancing their diversity [71, 72]. The implementation of these paths inevitably requires a teaching design based on methodologies such as tutoring, cooperative learning, and the metacognitive approach. The application of these paths seems simple to implement, but in practice, it is not, especially when stu-

Note: The chapter is a collaborative work between the two authors. For the purposes of the Italian research evaluation process, parts 1 to 3 are attributed to Rita Cersosimo , and parts 4 to 8 are attributed to Valentina Pennazio.

Corresponding author: Rita Cersosimo, DISFOR Department, School of Education, University of Genoa, Genoa, e-mail: rita.cersosimo@unige.it
Valentina Pennazio, DISFOR Department, School of Education, University of Genoa, Italy, e-mail: valentina.pennazio@unige.it

https://doi.org/10.1515/9783111352695-006

dents with special needs, such as those with autism spectrum disorder (from here on: "autism"), are present in the classroom. In these cases, teachers and parents of these students express various difficulties in creating inclusive learning paths and meaningful activities, managing interaction, and stimulating interest and participation. One of the most critical issues for teachers to address is the promotion of positive social interaction because the limited social skills of students with autism, both in verbal communication and in relationships regulated by socio-cultural norms, can lead to "isolation in the classroom," especially when the level of severity of the disorder is more marked [73].

This chapter presents a project funded by the Italian Foundation for Autism (FIA), a nonprofit organization. The first part of the chapter theoretically outlines the social and communicative profiles of individuals with autism, followed by a discussion on the use of social robots within this population. Research has demonstrated that this technology is effective in rehabilitation and educational interventions, aiding in the acquisition of social, communication, and academic skills. As described in the ICF [74], these tools can act as contextual facilitators by promoting accessibility to activities and full participation for students with disabilities. The second part of the chapter describes an innovative framework that combines robotics and virtual worlds to develop social skills in children and adolescents with autism. The objectives vary based on the age of the participants, ranging from emotion recognition and communication to maintaining communication at differing levels of complexity, and will involve the use of several robots. Individualized work with each participant aims to develop specific skills that can be transferred to the school context. In line with the principles set out by Salend [75], an additional goal of the project is to establish a permanent research and training laboratory for teachers at all school levels. This lab will equip teachers with the skills to design effective learning paths for students with autism, utilizing active teaching methodologies and robotic tools.

2 Social communication in autism

The diagnostic criteria for autism, as outlined in the DSM-5 [2], include "persistent deficits in social communication and social interaction" as a defining feature of this condition. Moving away from a medical perspective and adopting a pedagogical one, this implies that people with autism may require tailored support to navigate social environments from childhood into adulthood. While it is imperative for society to acknowledge the diversities inherent in all human beings, numerous research studies suggest that genuine inclusion stems from effective communication and mutual understanding among individuals [76, 77].

People with autism exhibit a heterogeneous range of language and social skills, with a significant portion – estimated to be between 25% and 30% – remaining nonverbal or possessing only minimal verbal abilities [78, 79]. Even among those who develop language skills, challenges in social communication persist [80, 81], encompassing both functional and pragmatic impairments. From early childhood, comprehension difficulties may emerge [80], and achieving speech fluency (i.e., using multiword combinations spontaneously, communicatively, and consistently) tends to be difficult and delayed [3, 4].

Engaging in conversation necessitates not only verbal skills but also social–emotional reciprocity, turn-taking, eye contact, gestures, and facial expressions [82, 83, 84, 85]. These components may pose challenges for individuals with autism [86], who often report difficulties in communication across various social contexts (see [5] for the significance of considering self-reports, where feasible). Research indicates that individuals with autism encounter challenges related to topic information, such as incorporating irrelevant detail and providing vague or unrelated responses; topic management, including inappropriate topic shifts and low rates of initiations; and reciprocity, manifesting as poor responsiveness and a limit in follow-up questions [87, 88].

Two primary areas in which social communication difficulties manifest have been identified [6]. The first area is the capacity for joint attention, which involves challenges in coordinating attention between individuals and objects. The second area is the ability for symbol use, which entails difficulties in learning conventional or shared meanings for symbols and is apparent in acquiring gestures, words, imitation, and play.

The ability for joint attention is fundamental for the development of language and is typically acquired by children during their first year of life [7, 89]. By the age of 9 months, infants actively observe others and learn to shift their gaze between people and objects to check if caregivers are attending to their focus of interest [8]. During this period, children also begin to use signals such as sounds, gestures, and other behaviors to achieve goals [9], while simultaneously developing the capacity to share emotional states with others and interpret them [90]. Children with autism exhibit differences in joint attention abilities, characterized by reduced frequency of gaze shifts, diminished duration of joint engagement, and challenges in tracking another individual's focus of attention by observing their gaze direction or responding to their pointing gestures [91, 92, 93]. Difficulties in joint attention are likely to have a ripple effect on language development because language acquisition often takes place in the context of caregivers modeling words that refer to objects and events that are being jointly attended to [94].

Symbol use is the second area of social communication difficulties in autism, and it is related to the nonverbal aspect of language; it involves the acquisition of a repertoire of conventional sounds and gestures to express intentions [6]. Between the ages of 1 and 2, children acquire the ability to symbolize, which involves representing one thing with another. This development is observable in their capacity to mimic new actions, engage in imaginative play with objects, and comprehend and use words to denote objects and events [95, 96]. Since comprehension involves understanding both nonverbal and verbal communicative signals used by others and determining meaning based on the context, the capacity to symbolize grows in parallel with achievements in speech perception. Children with autism exhibit specific deficits in acquiring conventional communication skills as well as more general deficits that affect cognitive and symbolic functioning [6]. Some individuals with autism lack the use of conventional gestures, such as showing and waving, as well as symbolic gestures like nodding and mimicking [95, 97, 98, 99, 100, 101]. Additionally, engaging in symbolic play, where objects are used for roles other than their intended purposes seems to pose challenges for individuals with autism [10].

Joint attention and symbol use are intertwined with another factor that might influence social communication skills and the processes of sharing attention, affect, and intentions: theory of mind. This term refers to the ability to attribute mental states to oneself and others [102, 103, 104]. Theory of mind emerges in preschoolers and continues to develop throughout childhood, becoming particularly important for one's social life (e.g., friendship; [105]). The role of theory of mind in communication is crucial, as it enables us to infer the thoughts and emotions of others during interaction [11]. Research evidence shows that theory of mind tends to be impaired in children with autism and that its delay persists through adolescence into adulthood, causing difficulties in social communication (see [102, 104, 106, 107] for reviews). This difficulty also has implications for the development of empathic skills, which encompass both cognitive and emotional dimensions. At the cognitive level, it involves adopting a conceptual perspective [108] and discerning the thoughts and emotions of others. At the emotional level, it influences the inclination to generate an appropriate emotional response to the other person's state [12], thereby impacting the formation of effective social interactions.

To sum up, communication involves verbal abilities, facial expressions, vocal tones, and body language to express one's internal feelings and intentions to others [109]. Successful interpersonal dialog requires individuals to interpret facial cues, vocal nuances, sustain eye contact, grasp the flow of interaction, refrain from excessive questioning, and infer underlying messages [110]. Previous research has highlighted the effectiveness of structured interventions designed to enhance conversational skills in individuals with autism (for a systematic review, refer to [111]; for a more recent literature review, see [112]). Among these interventions, those mediated by technology have shown particular promise [113]. Technology-based interventions are well-suited to meet the needs of individuals with autism due to their engaging nature, ability to reduce distractions, predictability, and adaptability to specific user characteristics [114, 115].

3 Why use robotics for autism?

Starting from the identification of the aforementioned difficulties, research in recent times has demonstrated the potential of using robotics to support the development of social communication and emotional skills in individuals with autism. Robots, due to their predictability, emotional simplicity, and flexibility (adaptable by the educational designer based on the needs of the person), can be employed to help individuals with autism in learning social, emotional, and imitative skills, transferring the acquired knowledge to interactions with human partners [116]. The use of a robotic platform is regarded by some scholars in the field [14] as an attempt to bridge the gap between the stable, predictable, and safe environment of a simple toy and the complex and unpredictable world of human communication and interaction [15]. The social behavior of individuals is indeed unpredictable and may appear threatening to a person with autism, while the use of a robot provides a simplified environment (e.g., in language or gestures) that can facilitate a gradual increase in the complexity of interaction, based on the needs and abilities of the individual.

Recent literature reviews have highlighted several key areas of social communication skills in individuals with autism where robots can be effectively employed. Drawing from the findings reported in Pennazio [117] and Alghamdi et al. [16], we categorized these areas into two main domains. The first encompasses the precursors of theory of mind, such as joint attention and symbol use, including imitation and other nonverbal abilities like gestures and eye contact. The second domain pertains to general social communication skills, encompassing both receptive and expressive verbal language, emotional skills, and the ability to understand the mental states of others, which more strictly pertains to theory of mind itself.

3.1 Development of Theory of Mind precursors

Regarding the precursors of theory of mind, numerous studies have sought to investigate the effects of a robot in directing the attention of children with autism and maintaining it consistently during interaction, whether the robot is in a static position or in motion. For instance, Mehmood et al. [118] demonstrated that the speech stimulus delivered by the robot attracts the attention of children with autism more rapidly compared to other visual and motion stimuli, as evidenced by the latency in shifting attention and the frequency of attention paid. Warren et al.'s study [17] pointed to an improvement in the focus and regulation of joint attention and a sustained focus toward the actions and proposals of the humanoid robot. Specifically, children with autism showed an increased ability to orient themselves toward the requests of the robotic system. Moreover, they not only exhibited greater attention to robotic systems compared to human caregivers but also demonstrated superior imitation performance during robotic interactions. Imitation was also explored in Conti et al. [18], where a task aimed at teaching gross motor imitation skills to children was used. Subsequent to the robot training, children with autism were successfully able to perform entirely new gross motor imitations.

Kumazaki et al. [119] conducted a study comparing the behavior of children with autism to that of typically developing children in a joint-attention-elicitation task. During the task, participants interacted with either a human or a robotic agent, both attempting to induce joint attention by alternating gazes between the child and images on the left and right sides for specific durations. Their findings indicate that robotic intervention is more effective in enhancing joint attention among children with autism compared to interventions involving human agents. Furthermore, following interaction with the robot, children demonstrated improved performance in joint attention tasks involving humans.

Other studies aimed at investigating the possibility of developing the ability of children with autism to maintain eye contact with a static or moving object have suggested that the use of robotic mediators in playful sessions of rehabilitative intervention is effective in achieving this goal. Simut et al. [19] confirmed through their studies that children with autism are able to maintain eye contact more effectively with a social robot compared to a human caregiver. Other studies have highlighted an increase in the time spent looking at the robot (in both static and dynamic positions) compared to that dedicated to a human partner [20].

3.2 Communication and emotional skills

Some studies have also highlighted how, for a person with autism, it is much easier to approach and interact with a robot companion than with human interlocutors, who are unpredictable in their responses. In fact, the robot can be programmed according to specific needs, thus creating predictable and emotionally reassuring relational situations [120]. Most importantly, the interaction with the robot has been found to increase social acceptability, that is, the willingness of the child with autism to engage in relationships first with the robotic mediator and then with the human partner [21, 22].

The use of robots has also proven beneficial in acquiring collaborative behaviors and their application in relationships with human partners [23]. However, the study conducted by Huskens et al. [121] on improving collaborative behavior (i.e., initiation of interaction, responses, and shared play) among children with autism and their siblings during play sessions mediated by the NAO robot and the use of Lego® did not yield statistically significant changes in this direction, although some indicators of improvement were observed. From the perspective of verbal communication skills at a social level, Dunst et al. [24–26] conducted a series of studies demonstrating the positive impact of work mediated by a socially interactive robot on increasing conversational turns between children with autism and their mothers, as well as on vocalization production.

Soares et al. [122] explored how a humanoid robot impacts the development of socioemotional skills in children with autism. They assessed the children's performance in in-game scenarios designed to improve their ability to recognize facial expressions. The study's key findings suggest that employing a humanoid robot effectively enhances socioemotional skills in interventions for children with autism, as evidenced by increased engagement and positive learning outcomes.

Engaging in activities with interactive humanoid robots offers individuals the opportunity to explore emotional states by recognizing emotions and the underlying factors influencing them [27, 28]. Moreover, it contributes to understanding the thoughts and beliefs of others, fostering the development of perspectives different from one's own. In essence, robotics holds promise for enhancing theory of mind skills [123]. Previous research has linked robot use to aspects of cognitive flexibility, enabling individuals to perceive events from varied perspectives and adapt their actions to different contexts, as well as to emotion recognition [124]. Notably, some studies have adopted a narrative-based approach that combines the use of social stories with robotics to develop theory of mind skills [29, 124, 125] Social stories are concise narratives that present social situations or concepts in a structured and meaningful manner [30]. Typically, they comprise descriptive sentences outlining the relevant social scenario, perspective sentences describing potential thoughts or feelings of others in that situation, and directive sentences offering guidance on appropriate responses or behaviors. These stories are often tailored to an individual's specific requirements and can encompass various topics, such as greetings, sharing, turn-taking, and coping with changes in routine. For instance, Vanderborght et al. [125] demonstrated that children with autism showed improved social performance when exposed to a robot conveying social stories compared to interactions with a human interlocutor.

3.3 The potential of humanoid social robots

The choice of robotic support based on its specific characteristics becomes fundamental, especially when working with individuals with autism. The robots that can be chosen come in different types: mobile ones with four wheels (e.g., [31, 32]), anthropomorphic dolls and puppets (e.g., [33, 126]), zoomorphic ones [127], and humanoid ones [128, 129]. However, the design of a robot might play a significant role in determining user acceptance, and not all robot designs are equally effective for this purpose [16]. According to van Straten et al. [130], children with autism demonstrated improved performance when interacting with doll-like robots with non-humanlike appearances compared to identical sessions with robots having humanlike facial features. Moreover, the study noted that children tended to avoid making eye contact with robots that had too humanlike characteristics. The use of robots whose appearance is distant from humans can trigger interesting interactions that stimulate creativity, also avoiding the occurrence of the uncanny valley effect [131], which is a decrease in interaction interest when the robot is unable to sustain the specific characteristics of the social model it refers to in its actions. Humanoid robots may then be preferred because their body characteristics are simplified (but still different) versions of human ones. At the same time, imitation and emotional skills learned through interactions with these robots are more easily transferable to interactions with other humans [34].

In the case of individuals with autism, humanoid robots with high interaction capabilities and responsiveness undoubtedly promise greater possibilities of generalization and chances of recognizing and imitating emotions. Cabibihan et al. [34], in their review, highlighted the great potential that the presence of certain specific robot characteristics can have in addressing the typical social, emotional, and communication difficulties of children with autism. First, robots need to be visually engaging for a person with autism to guarantee higher concentration spans [35, 132, 133]; at the same time, too many lights or colors should be avoided to prevent overstimulation [132, 134]. Particularly when working with children, robots should ideally be sized to match the child undergoing intervention [145]; this facilitates easier eye contact and mitigates any potential intimidation during interaction. Furthermore, children with autism are often attracted to moving objects, and research experiments have shown their preference for engaging with interactive robotic toys over passive ones [36, 132]. Modularity and adaptability emerge as key features that a robot should possess [34]; modularity enables the execution of diverse functionalities tailored to different individuals, while adaptability entails a progressive enhancement in the complexity of interactions [132].

In a recent analysis conducted by Alghamdi et al. [16], it was noted that the predominant choice of robot in studies involving individuals with autism between 2017 and 2022 was the NAO robot (Fig. 1).

This marks a notable shift from the findings reported in Cabibihan et al. [34], where the studies employed doll-like, cartoonlike, and pet-like robots. The NAO robot, manufactured by SoftBank Group Corporation, has shown promising results in facilitating various aspects such as social interactions, affectivity, intervention, and assisted teaching, as demonstrated by Robaczewski et al. [128] in their review. However, one possible drawback of the NAO robot is its limited ability to express emotions through facial expressions, except for changes in eye LED colors. In a study by Cohen et al. [37], the recognition rates of the NAO, which conveyed emotions through postures, were compared with those of the iCat robot, which utilized facial expressions. While the overall recognition rates were higher for the iCat compared to the NAO, there was no significant difference when each emotion was examined separately. This suggests that both robots are capable of expressing emotions effectively, and facial expression may not be a crucial component of emotional expression. However, for interventions specifically targeting emotion recognition, it may be interesting to explore newer social robots recently introduced to the market. These robots are designed to more clearly express emotions and social behaviors, thereby potentially enhancing the effectiveness of interventions aimed at improving emotion recognition skills and theory of mind in individuals with autism. This is why, for the part of our intervention targeting preschool and primary school children, we have chosen the Buddy and Navel robots, respectively; we left NAO for secondary school students due to its more adultlike appearance.

Fig. 1: NAO robot.

Buddy was developed by Blue Frog Robotics, a French start-up based in Paris ([135] Fig. 2). It has a screen as a face to display meaningful and cartoonlike expressions; it also has the ability to react to human presence and assess engagement, as well as caress sensors to react to user actions.

Navel was developed by Navel Robotics within a German Federal Ministry of Education and Research project [136]. It has the ability to show vivid facial expressions and micromovements (Fig. 3), aligns itself with sound sources, and if it detects a face, it establishes eye contact.

These robots respect the ideal characteristics outlined by Cabibihan et al. [34], since they are of a childlike size, they are visually engaging but with neutral colors, and their sophisticated functions are extremely modulable and adaptable to the users' needs. However, to the best of our knowledge, there are still no published studies on their use with people with autism.

Before we proceed with the discussion of our project in this chapter, it is important to offer a word of caution. The role of the robot primarily revolves around acting as a social mediator between individuals with autism and human interlocutors, regardless of whether they are adults or children [34, 137]. Therefore, it is crucial to recognize that the robotic system should never be seen as a replacement for human interaction, but rather as a typical mediator in educational interventions.

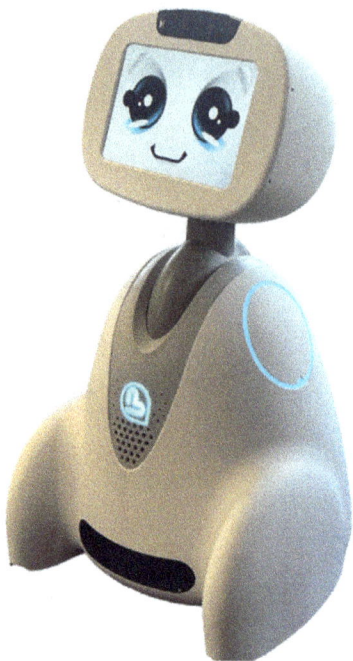

Fig. 2: Buddy robot.

Fig. 3: Navel robot (picture retrieved from Toussaint et al., 2023).

4 The research project

The presented research project is a continuation of a pilot study conducted in 2019–2020 in collaboration with the University of Macerata (Italy); the RobotiCSS Lab (Laboratory of Robotics for Cognitive and Social Sciences) of the Department of Human Sciences for Education "R. Massa", University of Milano-Bicocca (Italy); the social cooperative BES, Milan (Italy); and the Outpatient Clinic for Autism Spectrum Disorders in Adults of the ASST Santi Paolo e Carlo, Milan (Italy).

This pilot study involved an 8-year-old child diagnosed with autism, intending to verify whether the integration of two technological devices – a humanoid robot and an immersive virtual world – used as mediators for the presentation of social stories, was able to enhance the ability to recognize one's own and others' emotional states in children with autism. The idea behind the pilot study was to use the humanoid robot as a mediator to open a communication channel with the child. In a subsequent phase, a robot avatar in a virtual world was used as an expansion of the mediation function, with the possibility of presenting these stories (or different stories) in different environments, ensuring the generalization of the learned skills [138].

The project presented in this chapter therefore takes up the positive outcomes of the pilot study: opening a communication channel, focusing attention, encouraging communicative and verbal exchange, participation in carrying out activities, and generalization of skills. Starting from the results achieved in the pilot study, this project intends to extend the application of robotic supports from rehabilitation to school contexts, creating a network to structure inclusive learning paths for students with autism who attend various orders of school. The "enabling-the-use-of-tools" phase for children with autism will be carried out in a rehabilitation context where we will work on the recognition of emotions and their causes, maintaining communication, and practicing the ability to converse. In a school context, we will work on encouraging the application of the social skills previously learned in a triadic exchange situation between students with autism, the robot, and classmates.

The project aims to create a network between rehabilitation systems, families, and students to share the same educational intervention methodology, which is based on the application of social robotics for the achievement and development of lacking social skills. According to what is suggested by the literature, as presented in the previous paragraphs, social robots allow opening a communication channel with people with autism by channeling their attention and eye contact and triggering the implementation of new social behaviors [28, 38, 39, 40, 139, 140, 141, 142, 143]. Furthermore, robots, being programmable according to user needs, allow the creation of predictable and more reassuring relational situations from an emotional point of view compared to the unpredictability of human interlocutors [141, 144, 145]. The programmability of the robot allows the sensory information emitted to be varied, allowing, during the interaction, the modulation of stresses, expanding or reducing them based on the characteristics of the recipient. The robot is never intended as a substitute for the human being

[146] but as a social mediator that stands between the child and an adult or a peer [32]: the objective is always to encourage interaction with the human interlocutor. Humanoid robots can act as narrators of the story, capable of asking questions about the narrated situation; as actors in social situations carried out with human caregivers, which involve the presence of users in the role of passive spectators; or as actors together with the recipients, who must then demonstrate the understanding and internalization of what has been learned through the implementation of appropriate social actions [141]. The possibility of programming and customizing the characteristics of the robot allows the development of work sessions of increasing complexity, which are calibrated to the characteristics of the user.

Humanoid robots have therefore proven to be a valuable resource for encouraging the active participation of people with autism. However, as we have seen, they need to be carefully designed in their use and interaction characteristics to be accepted and used in a way that is valuable to their users. Starting from these considerations, this study aims, on the one hand, to create suitable educational ecosystems for social robotics characterized by interfaces, devices, and environments appropriately designed and integrated into the rehabilitation and educational paths of users with autism. This is intended to increase the perception of being in a particular context (which changes according to the social needs and goals that one wants to achieve) and to facilitate the learning and generalization of acquired social skills. On the other hand, the study aims to create a permanent training laboratory for teachers where they can learn to use these technological applications to promote the learning of students with autism.

5 Methodology and tools

To calibrate the intervention around the needs, desires, and real possibility of participation of children with autism, a user-centered design approach will be followed [41], namely a process in which attention is centered at every step of the design process on the characteristics of the user. This is essential to guarantee the usability of the product itself; the recipients of the project will therefore be actively involved, when possible, as they will be able to offer insights and ideas, but also stories and narratives, as well as feedback on the proposals put forward by the researchers. These suggestions will be useful for defining the work sessions to be implemented in the robot, the activities, the appearance, and methods of interaction with the robot, and the methods of evaluating the quality and effectiveness of learning paths mediated by robotic systems.

The study will follow both a top-down approach, in which researchers will suggest to therapists and teachers activities appropriate to the age and skills possessed by children with autism, to be implemented on the robot, and an approach focused on drawing information from the recipients themselves (therapists, teachers, pupils/students) concerning their needs. The robots used in the project will be Buddy, Navel, and NAO, as described in the previous paragraphs. The project will involve 16 participants with autism who attend the "PHILOS Pedagogical Academy" rehabilitation service (Genoa, Italy). Participants will be divided as follows: four children aged 4–5 years who attend nursery school; four children aged 6–10 years attending primary school; four boys aged 11–14 years who attend lower secondary school; four boys aged 15–18 who attend secondary school. Hence, this study lays emphasis on P-12 education.

6 Project articulation: phases and goals

The project will be divided into five phases, as described below.

6.1 Phase 1: emerging needs analysis in rehabilitative and school contexts

Through semi-structured interviews carried out by the PHILOS Pedagogical Academy with the families and, when possible, with the children and young people with autism who attend the center, the specific needs of the recipients will be identified (interaction needs, communication, participation). This information will be useful in the next phase to implement the activities on the robot and the work sessions.

At the same time, university researchers will administer questionnaires to teachers of the four school levels to detect training needs in terms of teaching methodologies and inclusive technologies known and used with pupils/students with autism. This analysis will serve to detect the emerging needs that teachers usually perceive at school when they have to work with a student with autism.

6.2 Phase 2: definition of therapeutic work sessions

Through a phase of discussion and co-planning between university researchers and PHILOS therapists, the activities will first be designed, defining the methodologies and strategies to be used and the role of the robot. Subsequently, these will be implemented on the devices through the creation of interfaces necessary to modify the perception and the degree of sensory stimulation produced, and the definition of the environmental scenarios within which the activity can be carried out.

In the case of nonverbal autism, a predominantly top-down approach will be followed; therefore, we will start from the interpretation of the needs expressed by the recipient. Some interventions and activities will be structured, which may be aimed at developing social skills, communication, or imitation, obviously depending on the school level. However, in the case of verbal autism and good interactive ability, the needs/preferences for setting up the robot's activities can be retrieved directly from the person involved.

For children with autism in kindergarten and the first classes of primary school, in a rehabilitation context, and also based on the type of need, we could hypothesize working with the robot to open the communication channel, stimulate imitation, and recognize emotions and the causes that produce them in association with the use of social stories. The robots to be used could be Buddy and Navel, considering their size and their capacity for emotional expression. For children with autism in lower and upper secondary schools, we could hypothesize, in a rehabilitation context, work on communication skills and the ability to hold a conversation; in this case, the NAO robot could be used.

6.3 Phase 3: practical experimentation of work sessions in a rehabilitative context

The experimentation of the work sessions will be preceded by the administration of tests to the recipients to detect the social and communication skills possessed before working with the robot. The same tests will then be administered at the end of the work sessions in a rehabilitation context to detect any changes that have occurred. For each participant, there are eight work sessions (8 weeks) lasting 15 min mediated by the robot, in which familiarization activities with the robot are planned; interactive activities differentiated based on the age of the participants, emotion recognition, communication management, and maintenance of conversation.

6.4 Phase 4: practical experimentation of work sessions in a school context

At the end of the experimental session in the rehabilitation context, work sessions will be organized in the school context and attended by the children who participated in the work sessions in the rehabilitation context. Robotic supports will be used as mediators in the learning activity within the classroom. First, work will be done in a small group composed of a student with autism, a robot, and a classmate, then work in a larger group composed of a student with autism, a robot, and two or three classmates. The teachers will plan the learning paths together with the university researchers. Also in this case, before the start of the work sessions, tests will be administered to evaluate the skills covered by the work sessions, and during the work sessions, observation grids will be used to monitor the progress of the activity.

6.5 Phase 5: structuring the training laboratory for teachers of all levels

Starting from the evidence drawn from the analysis of data extrapolated during the work phase in the rehabilitation and school context, the training path will be prepared, which will characterize the research and training laboratory and the exchange of good practices aimed at teachers of the four levels of school.

This laboratory will be aimed at disseminating practices of using robotics among teachers to structure inclusive learning paths and create networking with the rehabilitation services necessary for the preparation of shared working methodologies aimed at achieving objectives agreed upon by all professionals who work with individuals with autism.

7 Conclusions

Starting from the objectives that the project aims to achieve, we can describe the different potentialities connected to the project.

First, the proposal and monitoring of valid educational responses for the growth and education of persons with autism require an analysis of the real needs of both the recipients of the proposed interventions and the operators (therapists or teachers) who should implement the same routes.

Therefore, starting from the analysis of the different rehabilitation and school contexts allows us to have a realistic picture of the situation, not only to be able to grasp the needs of the people involved but also to plan specific responses. Thinking about the activities and calibrating the interventions to the specific needs of recipients with autism opens up numerous operational possibilities that allow us to accompany the recipients on a path of true growth and development.

The planning of rehabilitation and educational interventions shared between various professionals guides the choice of the most suitable methodologies and robotic supports for the specific situation and the specific needs of participants with autism. The union of experts from different fields (pedagogical, rehabilitation, technological) allows the creation of work that intervenes in different ways for the development of the same skills and competencies in the person with autism.

The project could also provide interesting food for thought from which to start in order to develop a more structured and stable path, both in a rehabilitation and scholastic context. The evidence drawn from the phases of the project will provide important data to continue research and improve rehabilitation and educational proposals for people with autism.

The creation of a permanent laboratory will allow a transfer of information and knowledge, which can then be brought to schools through the professionals who will attend the laboratory and the planned training. The possibility of making the laboratory permanent will make it possible to use technological tools that are difficult to find (such as robots) and will allow teachers to train by acquiring specific skills in the knowledge of autism, intervention methodologies, and the robotic tools that can be used.

Future research involves the extensive testing of this framework and its outcomes on specific social communication skills. Guidelines on how to reproduce this research design, adapted to the needs of each individual, will be identified and disseminated; this phase will involve professionals working in the rehabilitation field, but also teachers, who can use social robots to promote inclusion in the classroom setting.

References

[1]   Cole S, Horvath B, Chapman C, Deschedes C, Ebeling DG, Sprague J. Adapting curriculum & instruction in inclusive classrooms: A teacher's desk reference. 2nd ed. Bloomington, IN: The Center on Education and Lifelong Learning; 2000.

[2]   American Psychiatric Association. Diagnostic and statistical manual of mental disorders. 5th ed. 2013. https://doi.org/10.1176/appi.books.9780890425596

[3]   Baxter LC, Nespodzany A, Walsh MJM, Wood E, Smith CJ, Braden BB. The influence of age and autism on verbal fluency networks. Research in Autism Spectrum Disorders, 2019, 63, 52–62.

[4]   Begeer S, Wierda M, Scheeren AM, Teunisse JP, Koot HM, Geurts HM. Verbal fluency in children with autism spectrum disorders: Clustering and switching strategies. Autism, 2014, 18(8), 1014–1018.

[5]   Danés M, Botella J, Belinchón M. Validity of self-reports provided by people with autism spectrum disorder without intellectual disability: A meta-analysis. Annals of Psychology, 2023, 39(1), 88–99. https://doi.org/10.6018/analesps.509191

[6]   Charman T, Stone W. Social and communication development in autism spectrum disorders: Early identification, diagnosis, and intervention. Guilford Press; 2008.

[7]   Carpenter M, Nagell K, Tomasello M. Social cognition, joint attention, and communicative competence from 9 to 15 months of age. Monographs of the Society for Research in Child Development, 1998, 63, 1–143. (4, Serial No. 255).

[8]   Bakeman R, Adamson L. Coordinating attention to people and objects in mother-infant and peer-infant interaction. Child Development, 1984, 55, 1278–1289.

[9]   Bates E. The emergence of symbols: Cognition and communication in infancy. New York: Academic Press; 1979.

[10]   Gonzalez-Sala F, Gómez-Marí I, Tárraga-Mínguez R, Vicente-Carvajal A, Pastor-Cerezuela G. Symbolic play among children with autism spectrum disorder: A scoping review. Children, 2021, 8(9), 801.

[11]   Bosco FM, Tirassa M, Gabbatore I. Why pragmatics and theory of mind do not (completely) overlap. Frontiers in Psychology, 2018, 9, 1–8.

[12]   Davis M. Empathy: A social psychological approach. Boulder: Wes Tview Press; 1994.

[13]   Silver M, Oakes P. Evaluation of a new computer intervention to teach people with autism or Asperger syndrome to recognize and predict emotions in others. Autism, 2001, 5(3), 299–316.

[14]   Dautenhahn K, Nehaniv CL, Walters ML, Robins B, Kose-Bagci H, Assif N, Blow M. KASPAR–a minimally expressive humanoid robot for human-robot interaction research. Applied Bionics and Biomechanics, 2009, 6(3–4), 369–397.

[15] Cunha Costa SC. Affective robotics for socio-emotional development in children with autism spectrum disorders. Doctoral dissertation Braga, Portugal: Universidade di Minho; 2014.

[16] Alghamdi M, Alhakbani N, Al-Nafjan A. Assessing the potential of robotics technology for enhancing educational for children with autism spectrum disorder. Behavioral Sciences (Basel), 2023, 13(7), 598. https://doi.org/10.3390/bs13070598

[17] Warren ZE, Zheng Z, Swanson AR, Bekele E, Zhang L, Keshav NU, et al. Can robotic interaction improve joint attention skills?. Journal of Autism and Developmental Disorders, 2015, 45(11), 3726–3734.

[18] Conti D, Trubia G, Buono S, Di Nuovo S, Di Nuovo A. An empirical study on integrating a small humanoid robot to support the therapy of children with autism spectrum disorder and intellectual disability. Interaction Studies, 2021, 22(2), 177–211.

[19] Simut R, Vanderfaeillie J, Peca A, Van de Perre G, Vanderborght B. Children with autism spectrum disorders make a fruit salad with Probo, the social robot: An interaction study. Journal of Autism and Developmental Disorders, 2016, 46(4), 1134–1151.

[20] Esubalew B, Crittendon JA, Swanson A, Sarkar N, Warren ZE. Pilot clinical application of an adaptive robotic system for young children with autism. Autism: The International Journal of Research and Practice, 2014, 18(5), 598–608.

[21] De Graaf MMA, Ben Allouch S. Exploring influencing variables for the acceptance of social robots. Robotics and Autonomous Systems, 2013, 61, 1476–1486.

[22] Dunst CJ, Trivette CM, Prior J, Hamby DW, Embler DP. Judgments of the acceptability and importance of socially interactive robots for intervening with young children with disabilities. Social Robots Research Reports, 2013, 1, 1–5.

[23] Charron N, Lewis L, Craig M. A robotic therapy case study: Developing joint attention skills with a student on the autism spectrum. Journal of Educational Technology Systems, 2017, 46(1), 137–148.

[24] Dunst CJ, Prior J, Hamby DW, Trivette CM. Influences of a socially interactive robot on the affective behavior of young children with disabilities. Social Robots Research Reports, 2013a, 3, 1–10.

[25] Dunst CJ, Hamby DW, Trivette CM, Prior J, Derryberry G. Vocal production of young children with disabilities during child-robot interactions. Social Robots Research Reports, 2013b, 5, 1–7.

[26] Dunst CJ, Trivette CM, Hamby DW, Prior J, Derryberry G. Effects of child-robot interactions on the vocalization production of young children with disabilities. Social Robots Research Reports, 2013c, 4, 1–10.

[27] Barakova EI, Lourens T. Expressing and interpreting emotional movements in social games with robots. Personal and Ubiquitous Computing, 2010, 14, 457–467.

[28] Costa S, Lehmann H, Dautenhahn K, Robins B, Soares F. Using a humanoid robot to elicit body awareness and appropriate physical interaction in children with autism. International Journal of Social Robotics, 2014, 7(2), 265–278.

[29] Gillesen JC, Barakova EI, Huskens BE, Feijs LM. From training to robot behavior: Towards custom scenarios for robotics in training programs for autism. IEEE International Conference on Rehabilitation Robotics, 2011, 1–7.

[30] Gray C. Social Stories 10.0: The new defining criteria. Jenison Autism Journal, 2004, 15, 1–21.

[31] Dautenhahn K, Werry I. Towards interactive robots in autism therapy: Background, motivation and challenges. Pragmatics & Cognition, 2004, 12(1), 1–35.

[32] Ferrari E, Robins B, Dautenhahn K. Therapeutic and educational objectives in robot assisted play for children with autism. The 18th IEEE International Symposium on Robot and Human Interactive Communication, 2009, 108–114.

[33] Bulgarelli D, Bianquin N, Molina P. Children with cerebral palsy playing with mainstream robotic toys: Playfulness and environmental supportiveness. Frontiers in Psychology, 2018, 9, 398627.

[34] Cabibihan JJ, Javed H, Ang M, Aljunied SM. Why robots? A survey on the roles and benefits of social robots in the therapy of children with autism. International Journal of Social Robotics, 2013, 5(4), 593–618. https://doi.org/10.1007/s12369-013-0202-2

[35] Giullian N, Ricks D, Atherton A, Colton M, Goodrich M, Brinton B. Detailed requirements for robots in autism therapy. IEEE International Conference on Systems, Man and Cybernetics, 2010, 259–602.

[36] Dautenhahn K. Roles and functions of robots in human society: Implications from research in autism therapy. Robotica, 2003, 21(4), 443–452.

[37] Cohen I, Looije R, Neerincx MA. Child's perception of robot's emotions: Effects of platform, context and experience. International Journal of Social Robotics, 2014, 6, 507–518.

[38] Boucenna S, Narzisi A, Tilmont E, Muratori F, Pioggia G, Cohen D, Chetouani M. Interactive technologies for autistic children: A review. Cognitive Computation, 2014, 6, 1–19.

[39] Pennazio V. Robotics as a play-based learning tool: A review of the literature. International Journal of Innovation in Science and Mathematics Education, 2017, 25(1), 66–74.

[40] Robins B, Dautenhahn K, Dickerson P. From isolation to communication: A case study evaluation of robot assisted play for children with autism with a minimally expressive humanoid robot. Proceedings of the Second International Conferences on Advances in Computer-Human Interactions, 2009, 205–211.

[41] Dong H, Cassim J, Coleman R, Clarkson J. Design for inclusivity: A practical guide to accessible, innovative and user-centred design. Aldeshot, UK: Gower Publishing; 2008.

[42] Baron-Cohen S, Tager-Flusberg H, Cohen DJ, eds. Understanding other minds: Perspectives from developmental cognitive neuroscience. 2nd ed. Oxford: Oxford University Press; 2000.

[43] Fedeli L, Pennazio V. Instructional Design and 3D Virtual Worlds: A Focus on Social Abilities and Autism Spectrum Disorder. In: Handbook of research on teaching with virtual environments and AI. Hershey: IGI Global; 2021. pp. 444–460.

[44] Feil-Seifer D, Mataric M. Automated detection and classification of positive vs. negative robot interactions with children with autism using distance-based features. Proceedings of the 6th International Conference on Human-robot Interaction, 2011, 323–330.

[45] Kiuppis F, Hausstätter R. Inclusive Education for All: Developments in the UNESCO Education Sector, 2009–2019. In: Kiuppis F, Hausstätter R, eds. Inclusive education for all: Changes in legislation, policy and practice in the UNESCO education sector, 2009–2019. Leiden, The Netherlands: Brill Sense; 2020. pp. 1–13.

[46] Kose H, Koutsonikola K, Michalowski M, Nieman K, Montague C, Feil-Seifer D, Mataric M. Starting from scratch: Developing a fruitful human-robot relationship with a new social robot. The 15th IEEE International Symposium on Robot and Human Interactive Communication, 2006, 456–461.

[47] Lang R, Ramdoss S, Sigafoos J, Green VA, Van Der Meer L, Tostanoski A, et al. Assistive Technology for People with Autism Spectrum Disorders. In: Volkmar FR, Reichow B, McPartland J, eds. Adolescents and adults with autism spectrum disorders. New York: Springer; 2014. pp. 213–234.

[48] Lehmann H, Iacono I, Dautenhahn K, Marti P, Robins B. Robot companions for children with down syndrome: A case study. Interaction Studies, 2014, 15(1), 99–112.

[49] Leontyev AN. Activity, consciousness, and personality. Englewood Cliffs: Prentice-Hall; 1978.

[50] Lerner MD, Mikami AY, Levine K. Socio-dramatic affective-relational intervention for adolescents with Asperger syndrome & high functioning autism: Pilot study. Autism, 2011, 15(1), 21–42.

[51] Looije R, Neerincx MA, Cnossen F. Persuasive robotic assistant for health self-management of older adults: Design and evaluation of social behaviors. International Journal of Human Computer Studies, 2010, 68(6), 386–397.

[52] Lopes J, Da Silva C, Sá J, Oliveira G, Chaveiro N. Importance of ICT in the inclusion process of children with autism spectrum disorder: The case of Portugal. Education and Information Technology, 2018, 24, 517–543.

[53] Lord C, Rutter M, DiLavore PC, Risi S. Autism diagnostic observation schedule: ADOS. Los Angeles: Western Psychological Services; 2002.

[54] Lotter V. Epidemiology of autistic conditions in young children. Social Psychiatry, 1966, 1(3), 124–135.

[55] Loveland KA. Social affordances and interaction II: Autism and the affordances of the human environment. Ecological Psychology, 1991, 3(2), 99–119.

[56] Mesibov GB, Shea V, Schopler E. The TEACCH approach to autism spectrum disorders. Boston, MA: Springer US; 2005. https://doi.org/10.1007/b138381

[57] Milne E, Griffiths H. Visual perceptual processing in children and adults with autism: The importance of within-task comparisons. Behavioral and Brain Functions, 2007, 3, 29.

[58] Piper A, O'Brien E, Morris MR, Winograd T. SIDES: A cooperative tabletop computer game for social skills development. Proceedings of the 2006 20th Anniversary Conference on Computer Supported Cooperative Work, 2006, 1–10.

[59] Robins B, Dickerson P, Stribling P, Dautenhahn K. Robot-mediated joint attention in children with autism: A case study in robot-human interaction. Interaction Studies, 2004, 5(2), 161–198.

[60] Robins B, Dautenhahn K. Tactile interactions with a humanoid robot: Novel play scenario implementations with children with autism. International Journal of Social Robotics, 2014, 6(3), 397–415. https://doi.org/10.1007/s12369-014-0228-0

[61] Rutter M. Autistic children: Infancy to adulthood. Semin Psychiatry, 1970, 2, 435–450.

[62] Sabanović S, Chang WL, Bennett CC, Piatt JA, Hakken D. A robot of my own: Participatory design of socially assistive robots for independently living older adults diagnosed with depression. Proceedings of the ACM/IEEE International Conference on Human-robot Interaction, 2013, 106–107.

[63] Scassellati B, Admoni H, Mataric M. Robots for use in autism research. Annual Review of Biomedical Engineering, 2012, 14, 275–294.

[64] Spaulding TJ, Lerner MD, Gadow KD. Social skills training for children with intellectual disabilities: A review and call for future research. Behavior Modification, 2013, 37(6), 987–1025.

[65] Toth K, Munson J, Meltzoff AN, Dawson G. Early predictors of communication development in young children with autism spectrum disorder: Joint attention, imitation, and toy play. Journal of Autism and Developmental Disorders, 2006, 36(8), 993–1005.

[66] Vygotsky LS. Mind in society: The development of higher psychological processes. Cambridge, MA: Harvard University Press; 1978.

[67] Wainer J, Ferrari E, Dautenhahn K, Robins B. The effectiveness of using a robotics class to foster collaboration among groups of children with autism in an exploratory study. Personal and Ubiquitous Computing, 2010, 14, 445–455.

[68] Wood JJ, Drahota A, Sze KM, Har K, Chiu A, Langer DA. Cognitive behavioral therapy for anxiety in children with autism spectrum disorders: A randomized, controlled trial. Journal of Child Psychology and Psychiatry, 2009, 50(3), 224–234.

[69] Zwaigenbaum L, Bryson S, Lord C, Rogers S, Carter A, Carver L, et al. Clinical assessment and management of toddlers with suspected autism spectrum disorder: Insights from studies of high-risk infants. Pediatrics, 2009, 123(5), 1383–1391.

[70] Kochhar, C., West, L., & Taymans, J. (2000). Successful Inclusion: Strategies for a Shared Responsibility. Englewood Cliffs, N.J.: Prentice-Hall, Merrill Education Publishers.

[71] Kurtts, S. A. (2006). Universal Design for Learning in Inclusive Classrooms. Electronic Journal for Inclusive Education, 1 (10).

[72] Hall, T. E., Meyer, A., & Rose, D. H. (2012). Universal Design for Learning in the Classroom; Practical Applications. New York: The Guilford Press.

[73] Lemaignan, S., Newbutt, N., Rice, L., & Daly, J. (2024). "It's important to think of Pepper as a teaching aid or resource external to the classroom": A social robot in a school for autistic children. International Journal of Social Robotics, 16(6), 1083–1104.

[74] World Health Organization. (2001). International classification of functioning, disability and health: ICF. World Health Organization. https://iris.who.int/handle/10665/42407

[75] Salend, S. J. (2000). Strategies and resources to evaluate the impact of inclusion programs on students. Intervention in School and Clinic, 35(5), 264–289.

[76] Vaughn, S., Elbaum, B., & Boardman, A. G. (2001). The social functioning of students with learning disabilities: Implications for inclusion. Exceptionality, 9(1–2), 47–65.

[77] Hughes, C., Golas, M., Cosgriff, J., Brigham, N., Edwards, C., & Cashen, K. (2011). Effects of a social skills intervention among high school students with intellectual disabilities and autism and their general education peers. Research and Practice for Persons with Severe Disabilities, 36(1–2), 46–61.

[78] Pickles, A., Anderson, D. K., & Lord, C. (2014). Heterogeneity and plasticity in the development of language: a 17-year follow-up of children referred early for possible autism. Journal of child psychology and psychiatry, and allied disciplines, 55(12), 1354–1362. https://doi.org/10.1111/jcpp.12269

[79] Tager-Flusberg, H., & Kasari, C. (2013). Minimally verbal school-aged children with autism spectrum disorder: the neglected end of the spectrum. Autism research: official journal of the International Society for Autism Research, 6(6), 468–478. https://doi.org/10.1002/aur.1329

[80] Lord, C. & Paul, R. (1997). Language and communication in autism. In D.J. Cohen & F.R. Volkmar (Eds.), Handbook of autism and pervasive developmental disorders (2nd ed.). (pp. 195–225). New York: John Wiley and Sons.

[81] Wetherby, A. M., Watt, N., Morgan, L., & Shumway, S. (2007). Social communication profiles of children with autism spectrum disorders late in the second year of life. Journal of autism and developmental disorders, 37(5), 960–975. https://doi.org/10.1007/s10803-006-0237-4

[82] Heldner, M., & Edlund, J. (2010). Pauses, gaps and overlaps in conversations. Journal of Phonetics, 38 (4), 555–568. https://doi.org/10.1016/j.wocn.2010.08.002

[83] Levinson, S. C., & Torreira, F. (2015). Timing in turn-taking and its implications for processing models of language. Frontiers in Psychology, 6(731), 1–17. https://doi.org/10.3389/fpsyg. 2015.00731

[84] Mastroianni, A. M., Gilbert, D. T., Cooney, G., & Wilson, T. D. (2021). Do conversations end when people want them to? Proceedings of the National Academy of Sciences, USA, 118(10), 1–9. https://doi.org/10.1073/pnas.2011809118

[85] Stolk, A., Verhagen, L., & Toni, I. (2016). Conceptual alignment: How brains achieve mutual understanding. Trends in Cognitive Sciences, 20(3), 180–191. https://doi.org/10.1016/j.tics.2015.11.007

[86] Moody, C. T., & Laugeson, E. A. (2020). Social Skills Training in Autism Spectrum Disorder Across the Lifespan. Child and Adolescent Psychiatric Clinics of North America, 29(2), 359–371. https://doi.org/10.1016/j.chc.2019.11.001

[87] Paul, R., Orlovski, S. M., Marcinko, H. C., & Volkmar, F. (2009). Conversational behaviors in youth with high-functioning ASD and Asperger syndrome. Journal of Autism and Developmental Disorders, 39(1), 115–125. https://doi.org/10.1007/s10803008-0607-1

[88] Politis, Y., Clemente, I., Lim, Z., & Sung, C. (2023). The development of the conversation skills assessment tool. Autism & Developmental Language Impairments, 8, 23969415231196063. https://doi.org/10.1177/23969415231196063

[89] Wetherby, A., & Prizant, B. (2002). Communication and Symbolic Behavior Scales Developmental Profile – First Normed Edition. Baltimore: Brookes.

[90] Stern, D. (1985). The interpersonal world of the infant. New York: Basic Books.

[91] Sigman M, Mundy P, Sherman T, Ungerer J. Social interactions of autistic, mentally retarded, and normalchildren and their caregivers. Journal of Child Psychology and Psychiatry 1986;27:647–656. [PubMed: 3771681]

[92] Stone WL, Ousley OY, Yoder PJ, Hogan KL, Hepburn SL. Nonverbal communication in two- and three-year old children with autism. Journal of Autism and Developmental Disorders 1997;27:677–696. [PubMed: 9455728]

[93] Wetherby AM, Prizant BM, Hutchinson TA. Communicative, social/affective, and symbolic profiles of young children with autism and pervasive developmental disorders. American Journal of Speech-Language Pathology 1998; 7:79–91.

[94] McArthur, D., & Adamson, L. B. (1996). Joint attention in preverbal children: Autism and developmental language disorder. *Journal of autism and developmental disorders*, *26*(5), 481–496.

[95] Nadel, J., Aouka, N., Coulon, N., Gras-Vincendon, A., Canet, P., Fagard, J., & Bursztejn, C. (2011). Yes they can!: An approach to observational learning in low-functioning children with autism. *Autism*, 15 (4), 421–435. https://doi.org/10.1177/1362361310386508

[96] Wolfberg, P., DeWitt, M., Young, G. S., & Nguyen, T. (2015). Integrated play groups: promoting symbolic play and social engagement with typical peers in children with ASD across settings. *Journal of autism and developmental disorders*, *45*(3), 830–845. https://doi.org/10.1007/s10803-014-2245-0

[97] Loveland, K. A., & Landry, S. H. (1986). Joint attention and language in autism and developmental language delay. *Journal of autism and developmental disorders*, *16*(3), 335–349. https://doi.org/10. 1007/BF01531663

[98] McHale, S. M., Simeonsson, R. J., Marcus, L. M., & Olley, J. G. (1980). The social and symbolic quality of autistic children's communication. *Journal of Autism and Developmental Disorders*, *10*(3), 299–310.

[99] Stone, W. L., & Caro-Martinez, L. M. (1990). Naturalistic observations of spontaneous communication in autistic children. *Journal of autism and developmental disorders*, *20*(4), 437–453.

[100] Wetherby, A. M., Prizant, B. M., & Hutchinson, T. A. (1998). Communicative, social/affective, and symbolic profiles of young children with autism and pervasive developmental disorders. *American Journal of Speech-Language Pathology*, *7*(2), 79–91.

[101] Wolfberg, P., DeWitt, M., Young, G. S., & Nguyen, T. (2015). Integrated play groups: promoting symbolic play and social engagement with typical peers in children with ASD across settings. *Journal of autism and developmental disorders*, *45*(3), 830–845. https://doi.org/10.1007/s10803-014-2245-0

[102] Baron-Cohen, S., Campbell, R., Karmiloff-Smith, A., Grant, J., & Walker, J. (1995). Are children with autism blind to the mentalistic significance of the eyes?. *British Journal of Developmental Psychology*, *13*(4), 379–398.

[103] Leslie A.M. (1991), The theory of mind impairment in autism: Evidence for a modular mechanism of development? In A. Whiten (a cura di), *Natural theories of mind: Evolution, development and simulation of everyday mindreading*, Oxford, Blackwell, pp. 63–78.

[104] Peterson, C. C., & Wellman, H. M. (2019). Longitudinal Theory of Mind (ToM) Development From Preschool to Adolescence With and Without ToM Delay. *Child Development*, *90*(6), 1917–1934. https:// doi.org/10.1111/cdev.13064

[105] Fink, E., de Rosnay, M., Wierda, M., Koot, H. M., & Begeer, S. (2014). Brief report: accuracy and response time for the recognition of facial emotions in a large sample of children with autism spectrum disorders. *Journal of autism and developmental disorders*, *44*, 2363–2368.

[106] Happé, F. G. (1995). The role of age and verbal ability in the theory of mind task performance of subjects with autism. *Child development*, *66*(3), 843–855.

[107] Siegal, M., & Peterson, C. C. (2008). Language and theory of mind in atypically developing children: Evidence from studies of deafness, blindness, and autism. In C. Sharp, P. Fonagy, & I. Goodyer (Eds.), Social cognition and developmental psychopathology (pp. 81–112). Oxford University Press. https://doi.org/10.1093/med/9780198569183.003.0004

[108] Shantz C.U. (1983), Social Cognition. In P.H. Mussen (ed), *Handbook of Child Psychology*, (pp.495–555), New York, Wiley.

[109] Dols, J. M. F., & Russell, J. A. (Eds.). (2017). *The science of facial expression*. Oxford University Press.

[110] Politis, Y., Clemente, I., Lim, Z., & Sung, C. (2023). The development of the conversation skills assessment tool. *Autism & Developmental Language Impairments*, *8*, 23969415231196063. https://doi. org/10.1177/23969415231196063

[111] Hirvikoski, T., Jonsson, U., Halldner, L., Lundequist, A., De Schipper, E., Nordin, V., & Bölte, S. (2014). A Systematic Review of Social Communication and Interaction Interventions for Patients with Autism Spectrum Disorder. *Scandinavian Journal of Child and Adolescent Psychiatry and Psychology, 3* (3), 147–168. https://doi.org/10.21307/sjcapp-2015-016

[112] Moody, C. T., & Laugeson, E. A. (2020). Social Skills Training in Autism Spectrum Disorder Across the Lifespan. *Child and Adolescent Psychiatric Clinics of North America, 29*(2), 359–371. https://doi.org/10.1016/j.chc.2019.11.001

[113] Ramdoss, S., Lang, R., Mulloy, A., Franco, J., O'Reilly, M., Didden, R., & Lancioni, G. (2011). Use of Computer-Based Interventions to Teach Communication Skills to Children with Autism Spectrum Disorders: A Systematic Review. *Journal of Behavioral Education, 20*(1), 55–76. https://doi.org/10.1007/s10864-010-9112-7

[114] Moore, D., McGrath, P., & Thorpe, J. (2000). Computer-aided learning for people with autism: A framework for research and development. *Innovations in Education and Training International, 37,* 218–228.

[115] Panyan, M. (1984). Computer technology for autistic students. *Journal of Autism and Developmental Disorders, 14,* 375–382.

[116] Tapus, A., Maja, M., & Scassellatti, B. (2007). The grand challenges in socially assistive robotics. *IEEE Robotics and Automation Magazine, 14*(1), N-A.

[117] Pennazio, V. (2019). Robotica e sviluppo delle abilità sociali nell'autismo. Una review critica. *Mondo digitale, rivista di cultura informatica, 82,* 1–24.

[118] Mehmood, F., Mahzoon, H., Yoshikawa, Y., Ishiguro, H., Sadia, H., Ali, S., & Ayaz, Y. (2021). Attentional behavior of children with ASD in response to robotic agents. *IEEE Access, 9,* 31946-31955.

[119] Kumazaki H., Warren, Z., Muramatsu, T., Yoshikawa, Y., Matsumoto, Y., Miyao, M., Nakano, M., Mizushima, S., Wakita, Y., Ishiguro, H., Mimura, M., Minabe, Y. & Kikuchi, M. (2017). A pilot study for robot appearance preferences among high-functioning individuals with autism spectrum disorder: Implications for therapeutic use. *PLoS ONE, 12*(10).

[120] Robins, B., Dautenhahn, K., Boekhorst, R. T., & Billard, A. (2005). Robotic assistants in therapy and education of children with autism: can a small humanoid robot help encourage social interaction skills?. *Universal access in the information society, 4,* 105–120.

[121] Huskens, B., Palmen, A., Van der Werff, M., Lourens, T. & Barakova, E. I. (2015). Improving Collaborative Play between Children with Autism Spectrum Disorders and Their Siblings: The Effectiveness of a Robot-Mediated Intervention Based on Lego® Therapy. *Journal of Autism and Developmental Disorders,* 45 (11): 3746–3755.

[122] Soares, F. O., Costa, S. C., Santos, C. P., Pereira, A. P. S., Hiolle, A. R., & Silva, V. (2019). Socio-emotional development in high functioning children with Autism Spectrum Disorders using a humanoid robot. *Interaction Studies, 20*(2), 205–233.

[123] Pennazio, V., & Fedeli, L. (2019). A proposal to act on Theory of Mind by applying robotics and virtual worlds with children with ASD. *Journal of E-Learning and Knowledge Society,* Vol 15 No 2. https://doi.org/10.20368/1971-8829/1632

[124] Pop, C. A, Simut, R. E., Pintea S. & Vanderborght, B. (2013). Social Robots vs Computer Display: Does the Way Social Stories are Delivered Make a difference for Their Effectiveness on ASD Children? *Journal of Educational Computing Research,* 49(3): 381–401.

[125] Vanderborght B., Simut, R., Pop, J. C., Rusu, A. S., Pintea, S., Lefeber, D. & David, D. O. (2012). Using the social robot Probo as a social storytelling agent for children with ASD. *Interaction Studies* 13(3): 348–372.

[126] Kozima, H., Michalowski, M. P., & Nakagawa, C. (2009). Keepon: A playful robot for research, therapy, and entertainment. *International Journal of social robotics, 1,* 3–18.

[127] Stanton, C. M., Kahn Jr, P. H., Severson, R. L., Ruckert, J. H., & Gill, B. T. (2008). Robotic animals might aid in the social development of children with autism. *Proceedings of the 3rd ACM/IEEE international conference on Human robot interaction* (pp. 271–278).

[128] Robaczewski, A., Bouchard, J., Bouchard, K., & Gaboury, S. (2021). Socially assistive robots: The specific case of the NAO. *International Journal of Social Robotics*, *13*, 795–831.

[129] Pandey, A. K., & Gelin, R. (2018). A mass-produced sociable humanoid robot: Pepper: The first machine of its kind. *IEEE Robotics & Automation Magazine*, *25*(3), 40–48.

[130] van Straten, C. L., Smeekens, I., Barakova, E., Glennon, J., Buitelaar, J., & Chen, A. (2018). Effects of robots' intonation and bodily appearance on robot-mediated communicative treatment outcomes for children with autism spectrum disorder. *Personal and Ubiquitous Computing*, *22*, 379–390.

[131] Mori, M., MacDorman, K. F., & Kageki, N. (2012). The uncanny valley [from the field]. *IEEE Robotics & automation magazine*, *19*(2), 98–100.

[132] Michaud, F., Duquette, A., & Nadeau, I. (2003). Characteristics of mobile robotic toys for children with pervasive developmental disorders. In *SMC'03 Conference Proceedings. 2003 IEEE International Conference on Systems, Man and Cybernetics. Conference Theme-System Security and Assurance (Cat. No. 03CH37483)* (Vol. 3, pp. 2938–2943). IEEE.

[133] Robins, B., Otero, N., Ferrari, E., & Dautenhahn, K. (2007). Eliciting requirements for a robotic toy for children with autism-results from user panels. In *RO-MAN 2007-The 16th IEEE International Symposium on Robot and Human Interactive Communication* (pp. 101–106). IEEE.

[134] Hoa, T. D., & Cabibihan, J. J. (2012). Cute and soft: baby steps in designing robots for children with autism. In *Proceedings of the Workshop at SIGGRAPH Asia* (pp. 77–79).

[135] Milliez, G. (2018). Buddy: A Companion Robot for the Whole Family. *Companion of the 2018 ACM/ IEEE International Conference on Human-Robot Interaction*, 40–40. https://doi.org/10.1145/3173386. 3177839

[136] Toussaint, C., Schwarz, P. T., & Petermann, M. (2023). Navel – A social robot with verbal and nonverbal communication skills. *Extended Abstracts of the 2023 CHI Conference on Human Factors in Computing Systems*, 1–4. https://doi.org/10.1145/3544549.3583898

[137] Pennazio, V. (2019). Robotica e sviluppo delle abilità sociali nell'autismo. Una review critica. *Mondo digitale, rivista di cultura informatica*, *82*, 1–24.

[138] Pennazio V, Fedeli L, Datteri E., Crifaci G. (2020), Robotica e mondi virtuali per lo sviluppo delle abilità sociali nei bambini con autismo: una riflessione metodologica, Sistemi intelligenti, a. XXXII, n.1, ISSN1120-9550, pp. 139–154.

[139] Lytridis, C., Vrochidou, E., Chatzistamatis, S. & Kaburlasos V. (2019). Social engagement interaction games between children with Autism and humanoid robot NAO. In M. Graña J.M., López-Guede O., Etxaniz Á., Herrero J.A., Sáez H., Quintián E. & Corchado (eds.), *International Joint Conference SOCO'18-CISIS'18-ICEUTE'18. 562–570*, Cham: Springer.

[140] Pennazio, V. (2015), Disabilità, gioco e robotica: una ricerca nella scuola dell'infanzia. *TD – Tecnologie Didattiche*, 23(3), (pp.155–163).

[141] Pennazio, V. (2019). Robotica e sviluppo delle abilità sociali nell'autismo. Una review critica. *Mondo digitale, rivista di cultura informatica*, *82*, 1–24.

[142] Robins, B., Dautenhahn, K., Nehaniv, C.L., Mirza, N. A., François, D. & Olsson, L. (2005). Sustaining Interaction Dynamics and Engagement in Dyadic Child-Robot Interaction Kinesics: Lessons Learn from an Exploratory Study. *In Robot and Human Interactive Communication*, 2005. (pp.716–722), ROMAN, IEEE International Workshop, Nashville, USA.

[143] Scassellati, B., Boccanfuso, L., Huang, C. M., Mademtzi, M., Qin, M., Salomons, N., Ventola, P. & Shic, F. (2018). Improving social skills in children with ASD using a long-term, in-home social robot. *Science Robotics* 3, (pp. 1–9).

[144] Robins, B., Otero, N., Ferrari, E., & Dautenhahn, K. (2007). Eliciting requirements for a robotic toy for children with autism-results from user panels. In *RO-MAN 2007-The 16th IEEE International Symposium on Robot and Human Interactive Communication* (pp. 101–106). IEEE.

[145] Robins, B., Dautenhahn, K., Nehaniv, C.L., Mirza, N. A., François, D. & Olsson, L. (2005). Sustaining Interaction Dynamics and Engagement in Dyadic Child-Robot Interaction Kinesics: Lessons Learn from an Exploratory Study. *In Robot and Human Interactive Communication*, 2005. (pp.716–722), ROMAN, IEEE International Workshop, Nashville, USA.

[146] Lytridis, C., Vrochidou, E., Chatzistamatis, S. & Kaburlasos V. (2019). Social engagement interaction games between children with Autism and humanoid robot NAO. In M. Graña J.M., López-Guede O., Etxaniz Á., Herrero J.A., Sáez H., Quintián E. & Corchado (eds.), *International Joint Conference SOCO'18-CISIS'18-ICEUTE'18. 562–570*, Cham: Springer.

Bettina Trixler and Henriette Pusztafalvi

Robots for autism-specific education: opportunities and barriers to the use of technical devices

Abstract: People with autism spectrum disorders have varying degrees of difficulty in social interactions. The worsening of symptoms and the emergence of challenging behaviors can be a barrier to learning, which is provided by interpersonal relationships. The question may be raised as to the impact of the reduction of the infinite number of unpredictable social outputs between persons on the condition of people with autism. Technological devices and robots can create an environment that can develop the individual's abilities in a personalized way, limiting distracting stimuli in the environment but maintaining interest. The use of technology can be a good complementary tool to therapy and teaching, but it requires careful pedagogical work to structure learning situations appropriately. Based on current knowledge, there are positive examples of reducing the symptoms of the condition, but more in-depth research on the topic is warranted to evaluate interventions over time, with a wider range of target populations.

Keywords: Autism spectrum disorders, skills, robots, use, symptoms, impact

1 Introduction

Autism spectrum disorder is a neurodevelopmental disorder associated with atypical manifestations of reciprocal social interactions and communication, stereotyped repetitive activities, interests, and behaviors. Symptoms are present early in development but can become more pronounced with age as social expectations change. The symptoms have a significant impact on the individual's abilities [1]. Autism is a highly heterogeneous condition, with variable symptoms influenced by language abilities, intelligence, comorbidities, neuropsychological underpinnings, etiology [2], age, environmental factors, autism severity, and personality [3]. Evidence shows that the quality of life for people with autism spectrum disorder and their families can be significantly improved through appropriate lifelong evidence-based educational approaches [4]. Autism affects

---

Corresponding author: Bettina Trixler, Doctoral School of Health Sciences, Faculty of Health Sciences, University of Pécs, Pécs, Hungary, e-mail: trixler.bettina@gmail.com
Henriette Pusztafalvi, Department of Health Promotion and Public Health, Faculty of Health Sciences, University of Pécs, Hungary

https://doi.org/10.1515/9783111352695-007

1 in 100 children according to the World Health Organization [5], while other sources suggest that 1 in 54 children are on the spectrum [6].

The chapter on autism and the use of robots was created to summarize the impact on symptoms and to include the limitations and benefits of their use. There are countless studies on the subject, and a compilation and interpretation of researchers' knowledge is therefore warranted, also due to the increasing number of cases. We are living in a decade where the utilization of technology is part of our everyday lives and its benefits should be exploited to develop skills. The chapter emphasizes the impact of the use of robots on symptoms, the relevance of their application, and the importance of the knowledge of professionals. The remainder of the chapter presents the academic relevance of the topic focusing on robot-assisted technologies (Section 2), the influence of robots on autism symptoms (Section 3), their applicability (Section 4), and the role of professional (Section 5). Additionally, it further discusses the role of robots in supporting autism-specific education (Section 6) and provides conclusive remarks (Section 7).

2 Robot-assisted technologies

In the last decades, robot-assisted technologies have undergone significant changes and have been investigated for their importance in the development of socio-communication skills and the regulation of repetitive, stereotyped interests, activities, and behaviors [7]. Technology is a way for teachers to adapt the teaching process to the learning styles and interests of students. Socially assistive robots can interact with people with autism in a personalized way to support skill development in an accessible and affordable way [8]. The research found evidence that the majority of people with autism have a natural interest in technology and a good propensity to use technology and learn from computers [9]. A range of interventions have been developed for people with an autism diagnosis to improve their cognitive abilities and daily living skills and increase community participation and interaction while reducing symptoms. For example, the application of technology has been used as an adjunct in therapy sessions. As the number of people with autism increases, there is a need to support them with innovations that meet societal needs and improve the effectiveness of therapy [10]. However, to date, few professional development programs specifically train teachers in the use of robots in the classroom and there is limited information available on the expectations of teachers of students with autism regarding the use of robots and the forms of support they would need based on their experiences [11].

3 The impact of robots on autism symptoms

People with autism may have difficulty interpreting other people's body language, using eye contact, or understanding other people's emotions [12]. In terms of motivation and engagement in learning, robots can play an important role in learning the skills needed for cognitive and social development. They often have more complex needs, have other diagnoses associated with their autism condition, and require much more specialized support and intensive development [13]. New technologies can help to better understand the condition of autism and help to develop skills. The tools should be introduced in a differentiated, personalized way, after a thorough assessment, and adapted to the individual needs of the child. Results report short-term positive results in the behavior of individuals with autism, but a summary of long-term experience is essential to justify the applicability of the procedure [14].

The outcome shows that technical tools can stimulate motivation, create a predictable environment, and allow autonomous task performance [15]. Robots are also used to reduce maladaptive behaviors. Particular attention should also be paid to establishing that interaction with robots does not reinforce compulsive behaviors or allow room for social isolation [16]. A reduction in challenging behaviors was also observed with the use of technology, although repetitive and stereotypical behaviors were observed in some educational sessions [17]. Communication functions and means, language and daily life skills, imitation and turn-taking, as well as play, attention, speech or vocalization, understanding cause-and-effect relationships, and coping skills may improve by robot technologies [18].

An evaluation of the creation and applicability of assistive robots has begun. Although people with autism show a high propensity to use the technology, there are many negative examples in the literature, and caution is advised regarding patterns of use, as their impact on emotional and motor functions is questionable [19]. Research has also raised the question of how dependent the interaction with the robot is, and whether the child will be able to generalize what they have learned to other situations and people. Assessing and understanding the preferences of the person concerned is the basis for individualized intervention. Robot-assisted therapy can be a tool for developing cognitive flexibility, problem-solving, adaptive skills, and understanding social rules [20]. In any case, it helps to practice skills such as joint attention, eye contact, imitation skills, initiating and maintaining interactions, and playing together. It can also stimulate the development of their interests and their broad knowledge of the world during effective practice [21].

New technologies can provide a safe and reliable environment for people with autism in line with stimulus and task control [22], and technological advancements can also help them to learn useful everyday skills such as financial and environmental awareness, acquiring health literacy, or use as a reward [23]. However, people diagnosed with autism do not always respond positively to robot manifestations, which reinforce the heterogeneous nature of the population and the need for personalized therapy [24]. After all, the relative lack of data on autism and robotics from different countries around the world reinforces the need for more global perspectives on robot interactions, intervention design and process, and cultural influences [25].

4 The relevance of the application

Robots are designed to perform functions and tasks that improve the quality of life of the person concerned, increase their independence and autonomy from their environment, and reduce the burden on those around them. These tools create a more predictable environment for learning in a safe environment. The results are still mixed, but encouraging as they have been tested across a wide range of skills. In all cases, individualized interventions can help to eliminate negative effects. One of the fundamental aspects of designing robots is to avoid distraction. In terms of results, the overall effects of introducing robots in therapies have been positive. Researchers have seen positive effects on children's behavior, attention, and learning. There are also questions about how artificial intelligence and machine learning can improve the adaptability of robots to the changing needs of children with autism. Another question is how robots and human therapists can collaborate during therapies: could they act as assistants in the implementation of therapy programs, and could they collect comprehensive data to assess the child's development? Their applicability in the classroom has also been studied in limited numbers and with limited methods to effectively support children with autism. In addition, robotic remote monitoring and telehealth can be explored to provide remote and personalized therapy. The formal adoption of robots in a therapeutic setting could trigger investigations into privacy, ethical, and societal issues, which will be essential to explore. Consideration should also be given to the long-term evaluation of the impact of robotic interventions on the development and well-being of children with autism [14].

The possible future of screening is foreshadowed by social robots [26]. This wide range of presentations encourages the achievement of therapeutic goals. They can help with symptom management, different types of interactions, and finding employment as an adult. Robot-assisted autism therapy is a relatively new field in the expansion and extension of treatments, but they are all designed to achieve one goal: to support the improvement of the quality of life of children with autism and their families. The level and variability of interactions with robots are not expected to reach the level of rich interactions with humans shortly, but their use as a therapeutic tool could prove to be very productive in building relationships with other humans. Although currently unproven, human-like robots may also provide more opportunities for the generalization of skills. A drawback of using this new technique is that robots may also be less attractive to children with autism [27]. In this respect, factors such as appearance, movement, or clothing become important. The affinity of autistic individuals for the robot may also be influenced by the age, gender, and intelligence level of the individual [28].

Criticisms of the research that has been done on this topic are as follows: lack of a control group, low sample size, and assessment in a clinically controlled, structured environment. Testing and understanding theories in practice are therefore complicated by several factors. Studies may show trends, but it is not possible to generalize these results to the population. Future studies are essential to resolve differences of opinion between researchers. Strategies to measure the generalizability of educational outcomes are designed to monitor effectiveness [29].

5 Professionals

The diversity of technological advances also requires a high level of preparedness and adaptation on the part of professionals. The integration of these tools into the day-to-day development of public institutions also requires significant infrastructure, resources, and protocol. The attitudes and skills of professionals also need to be further explored to develop effective interventions [30]. Based on the results of interventions to date, robotic therapy appears promising, with caregivers reporting positive changes in social participation in addition to short-term improvements in skills [31]. However, the lack of evaluation of long-term outcomes rightly generates uncertainty among professionals about the use of robots [32], even if robot support can reduce workload by automating repetitive interventions [33]. Professionals' concerns are significant in terms of the acquisition of autism-related skills, the lack of which may be reflected in the quality of teaching practice. Where there are accessible, credible resources available to professionals about the care of people with autism, they are more likely to be equipped to develop individualized programs. As professionals' knowledge increases, their confidence also improves. Unfortunately, to date, few professional development programs specifically prepare teachers to use robots in the classroom. There is also limited information available on teachers' perceptions and expectations regarding the introduction and usage [11]. As the vast majority of studies have been conducted in clinical settings, further research is needed to explore them in a broader context [34]. In the near future, the development of a new generation of tools with high levels of mobility, availability, safety, and acceptability for improving the complex triadic interaction among teachers, children, and robots will certainly be warranted [35].

Overall, technological advances provide professionals with many opportunities to understand, learn about, and develop people with autism spectrum disorders. Nevertheless, we must also be careful to avoid several negative impacts. This is particularly important given the wide spectrum and heterogeneity of symptoms. The development of adapted tools also requires interdisciplinary collaboration to follow specific needs. The use of robots in concrete practical situations is still difficult because it requires a deep understanding of the professionals who help them. It is therefore essential to comprehend the views that may be barriers or advantages in the introduction of new technologies. Understanding the impact of the intervention is a cardinal issue for the transfer of techniques from controlled, supervised trials and situations to real life. There is a need for a broader knowledge of the usefulness of robots due to the specific requirements of autistic people. Professionals tend to work with people with different conditions and symptoms at the same time, so introducing new technology to groups of learners with special needs can be challenging. In this respect, the current skill level of the teachers and the adaptability of the robot may also create a significant barrier [36].

Careful selection of situations where electronic tools can provide a real solution and do not hinder the development of skills is needed. Family members know their child's needs best; therefore, collaboration can be very effective. Understanding the child's wider environment and taking advantage of opportunities is also necessary to inform interventions. Parents are mostly accepting the use of robots, seeing them as a way to complement the work of professionals [37].

The wide range of behaviors, preferences, and characteristics of learners with autism requires fine-tuning robots with appropriate technical background knowledge. Demonstrating feasibility and effectiveness can be difficult, as the robot can be used in almost any way; therefore, the learning gains need to be measurable. The needs of adult users who may operate robotic systems should be addressed in addition to the needs of child users. Beyond the development of controlled robot experiments and special educational programs, it would be necessary for the representatives of the individual professions to actively work together for effective and safe use. Carefully outline the specific use cases and circumstances in which robots are beneficial. The development of a training program is recommended to regulate the proper integration of robots in therapy [38].

6 Discussion

In clinical practice, in diagnosis and therapy, new, modern methods can offer several practical solutions [39]. Research findings show significant effects of such learning on social skills, emotion regulation and recognition, and communication skills [40]. The emergence of robots may influence their adoption [41], and it is also necessary to consider and weigh up the applicability and functionality of different types of robots [42]. In a learning environment, robots can support individual engagement and skill development by providing reliable and repeatable demonstrations and interactions. Their application can allow them to complement and improve on previous knowledge and achievements. They can measurably detect skill development [43] by providing predictable, consistent behavior and a limited stimulus environment [44]. Socially assistive robots can interact autonomously or semi-autonomously with autistic people to support their daily activities. They have the potential to complement the work of educators and therapists during the provision of personalized, accessible, and affordable interventions and support [45].

Robots can be a motivating factor for children [46], but they can also be used in a wide range of applications, from job interview preparation [47] to healthcare [48]. Although data show a positive impact on the development of people with mental health disorders, technical problems, practitioner attitudes, and inappropriate environments are key factors affecting the success of interventions [49]. Examples in the literature show favorable support for the learning of relaxation techniques, preventing the increase of irritation, developing self-regulation skills, and reducing disruptive behaviors [50]. It has also been shown to affect physical activity stimulation [51], but it is necessary to assess the stimuli to which individuals with autism are sensitive before they are involved in robot-assisted therapy and it fails to produce results [52]. The degree of disability is expected to influence the case-by-case usefulness of robots [53].

7 Conclusion

The results of international research clearly outline the benefits of using robotic technology, but several uncertainties and questions remain about the effects of long-term intervention, whether follow-up is a priority in trials, and whether the procedure is effective in the population as a whole. It appears that robots can provide effective support in interpreting and understanding social cues and can also perform specific educational tasks. However, the tools should be chosen carefully, adapted to the individual needs, and the learning situations should be structured so as not to produce unwanted effects. The major technological advances of the twenty-first century can also benefit the abilities of people with autism, including through their motivational nature. The simulation of situations can facilitate the practical acquisition of emotional and social cues, providing a sense of security in the learning process. However, it should not be forgotten that it may lead to dependency and that technological errors may make effective learning more difficult. It is also important to keep in mind that the complexity of situations cannot be modeled faithfully and completely. It may also be advisable to compare the types of robots and use various measurement methods in research to measure and evaluate changes in skills more accurately.

References

[1] American Psychiatric Association. Neurodevelopmental Disorders. In: Diagnostic and statistical manual of mental disorders. 5th ed. text rev. Washington, DC: APA; 2022. pp. 211–232.

[2] Francis K, Karantanos G, Al-Ozairi A, AlKhadhari S. Prevention in autism spectrum disorder: A lifelong focused approach. Brain Sciences, 2011, 11(2), 151. doi: 10.3390/brainsci11020151

[3] Stefanik K, Prekop Cs. Autizmus Spektrum Zavarok. In: Balázs J, Miklósi M, eds. A gyermek- és ifjúkor pszichés zavarainak tankönyve. Budapest, Hungary: Semmelweis Kiadó és Multimédia Stúdió; 2015 pp. 61–67.

[4] Posar A, Visconti P. Long-term outcome of autism spectrum disorder. Türk Pediatri Arşivi, 2019 Dec 25, 54(4), 207–212. doi: 10.14744/TurkPediatriArs.2019.16768.

[5] World Health Organization. Autism. (Accessed: November 15, 2023, at https://www.who.int/news-room/fact-sheets/detail/autism-spectrum-disorders.)

[6] Maenner MJ, Shaw KA, Baio J, et al. Prevalence of autism spectrum disorder among children aged 8 years – Autism and developmental disabilities monitoring network, 11 sites, United States, 2016. MMWR. Surveillance Summaries, 2020, 69(4), 1.

[7] Alghamdi M, Alhakbani N, Al-Nafjan A. Assessing the potential of robotics technology for enhancing education for children with autism spectrum disorder. Behavioral Sciences, 2023, 13(7), 598. https://doi.org/10.3390/bs13070598

[8] Clabaugh C, Matari'c M, Escaping Oz. Autonomy in socially assistive robotics. Annual Review of Control, Robotics, and Autonomous Systems, 2019, 2, 33–61.

[9] Begum M, Serna RW, Yanco HA. Are robots ready to deliver Autism interventions? A comprehensive review. International Journal of Social Robotics, 2016, 8, 157–181. https://doi.org/10.1007/s12369-016-0346-y

[10] Alabdulkareem A, Alhakbani N, Al-Nafjan A. A systematic review of research on robot-assisted therapy for children with autism. Sensors (Basel), 2022 Jan 26, 22(3), 944. doi: 10.3390/s22030944

[11] Sulaimani MF. Autism and technology: Investigating elementary teachers' perceptions regarding technology used with students with autism. International Journal of Special Education, 2017, 32, 586–599.

[12] Lord C, Bishop SL. Recent advances in autism research as reflected in DSM-5 criteria for autism spectrum disorder. Annual Review of Clinical Psychology, 2015, 11, 53–70. https://doi.org/10.1146/annurev-clinpsy-032814-112745

[13] Desideri L, Negrini M, Malavasi M, et al. Using a humanoid robot as a complement to interventions for children with autism spectrum disorder: A pilot study. Advances in Neurodevelopmental Disorders, 2018, 2, 273–285. https://doi.org/10.1007/s41252-018-0066-4

[14] Gómez-Espinosa A, Moreno JC, Pérez-de la Cruz S. Assisted robots in therapies for children with autism in early childhood. Sensors, 2024, 24(5), 1503. https://doi.org/10.3390/s24051503

[15] Dubois-Sage M, Jacquet B, Jamet F, Baratgin J. People with autism spectrum disorder could interact more easily with a robot than with a human: reasons and limits. Behavioral Sciences, 2024, 14(2), 131. https://doi.org/10.3390/bs14020131

[16] Diehl JJ, Schmitt LM, Villano M, Crowell CR. The clinical use of robots for individuals with autism spectrum disorders: A critical review. Research in Autism Spectrum Disorders, 2012 Jan, 6(1), 249–262. doi: 10.1016/j.rasd.2011.05.006.

[17] Gkiolnta E, Zygopoulou M, Syriopoulou-Delli CK. Robot programming for a child with autism spectrum disorder: A pilot study. International Journal of Developmental Disabilities, 2023 May 17, 69(3), 424–431. doi: 10.1080/20473869.2023.2194568.

[18] Zorcec T, Ilijoski B, Simlesa S, et al. Enriching human-robot interaction with mobile app in interventions of children with autism spectrum disorder. Pril (Makedon Akad Nauk Umet Odd Med Nauki), 2021 Oct 26, 42(2), 51–59. doi: 10.2478/prilozi-2021-0021.

[19]   Kouroupa A, Laws KR, Irvine K, Mengoni SE, Baird A, Sharma S. The use of social robots with children and young people on the autism spectrum: A systematic review and meta-analysis. PLoS ONE, 2022, 17, e0269800. doi: 10.1371/journal.pone.0269800.

[20]   Zhang Y, Song W, Tan Z, et al. Could social robots facilitate children with autism spectrum disorders in learning distrust and deception?. Computers in Human Behavior, 2019, 98, 140–149. doi: 10.1016/j.chb.2019.04.008.

[21]   van den Berk-smeekens I, van Dongen-boomsma M, De Korte MW, et al. Adherence and acceptability of a robot-assisted pivotal response treatment protocol for children with autism spectrum disorder. Scientific Reports, 2020. May 15, 10(1), 1–1.

[22]   Robins B, Dautenhahn K. Tactile interactions with a humanoid robot: Novel play scenario implementations with children with autism. International Journal of Social Robotics, 2014, 6, 397–415.

[23]   Valencia K, Rusu C, Quiñones D, Jamet E. The impact of technology on people with autism spectrum disorder: A systematic literature review. Sensors (Basel), 2019 Oct 16, 19(20), 4485. doi: 10.3390/s19204485.

[24]   Feil-Seifer D, Mataric M. Automated detection and classification of positive vs. negative robot interactions with children with autism using distance-based features. Proceedings of the 6th International Conference on Human-Robot Interaction – HRI '11; 2011. doi: 10.1145/1957656.1957785.

[25]   Hashim R, Yussof H. Humanizing humanoids towards social inclusiveness for children with autism. Procedia Computer Science, 2017. Jan 1, 105, 359–364.

[26]   So WC, Wong E, Ng W, et al. Seeing through a robot's eyes: A cross-sectional exploratory study in developing a robotic screening technology for autism. Autism Research, 2024 Feb, 17(2), 366–380. doi: 10.1002/aur.3087.

[27]   Ricks DJ, Colton MB. Trends and considerations in robot-assisted autism therapy. 2010 IEEE International Conference on Robotics and Automation, doi: 10.1109/robot.2010.5509327.

[28]   Kumazaki H, Muramatsu T, Yoshikawa Y, et al. Optimal robot for intervention for individuals with autism spectrum disorders. Psychiatry and Clinical Neurosciences, 2020 Nov, 74(11), 581–586. doi: 10.1111/pcn.13132.

[29]   Santos L, Annunziata S, Geminiani A, et al. Applications of robotics for autism spectrum disorder: A scoping review. Review Journal of Autism and Developmental Disorders, 2023. https://doi.org/10.1007/s40489-023-00402-5

[30]   Conti D, Di Nuovo S, Buono S, et al. Robots in education and care of children with developmental disabilities: A study on acceptance by experienced and future professionals. International Journal of Social Robotics, 2017, 9, 51–62. https://doi.org/10.1007/s12369-016-0359-6

[31]   Chung EY, Kuen-Fung Sin K, Chow DH. Effectiveness of robotic intervention on improving social development and participation of children with autism spectrum disorder – A randomised controlled trial. Journal of Autism and Developmental Disorders, 2024 Jan 17, doi: 10.1007/s10803-024-06236-2.

[32]   Takata K, Yoshikawa Y, Muramatsu T, et al. Social skills training using multiple humanoid robots for individuals with autism spectrum conditions. Frontiers in Psychiatry, 2023 Jul 19, 14, 1168837. doi: 10.3389/fpsyt.2023.1168837.

[33]   Vagnetti R, Di Nuovo A, Mazza M, et al. Social robots: A promising tool to support people with autism. A systematic review of recent research and critical analysis from the clinical perspective. Review Journal of Autism and Developmental Disorders, 2024. https://doi.org/10.1007/s40489-024-00434-5

[34]   Kewalramani S, Allen KA, Leif E, Ng A. A scoping review of the use of robotics technologies for supporting social-emotional learning in children with autism. Journal of Autism and Developmental Disorders, 2023 Nov 28, doi: 10.1007/s10803-023-06193-2.

[35] Puglisi A, Caprì T, Pignolo L, et al. Social humanoid robots for children with autism spectrum disorders: a review of modalities, indications, and pitfalls. Children (Basel), 2022 Jun 25, 9(7), 953. doi: 10.3390/children9070953.

[36] Hughes-Roberts T, Brown D. Implementing a Robot-Based Pedagogy in the Classroom: Initial Results from Stakeholder Interviews. In: 2015 international conference on interactive technologies and games. Nottingham, UK; 2015. pp. 49–54. doi: 10.1109/itag.2015.18.

[37] Coeckelbergh M, Pop C, Simut R, Peca A, Pintea S, David D, Vanderborght B. A survey of expectations about the role of robots in robot-assisted therapy for children with ASD: Ethical acceptability, trust, sociability, appearance, and attachment. Science and Engineering Ethics, 2016 Feb, 22(1), 47–65. doi: 10.1007/s11948-015-9649-x.

[38] Huijnen CAGJ, Lexis MAS, Jansens R, De Witte LP. Mapping robots to therapy and educational objectives for children with autism spectrum disorder. Journal of Autism and Developmental Disorders, 2016, 46(6), 2100–2114. doi: 10.1007/s10803-016-2740-6.

[39] Chevalier P, Kompatsiari K, Ciardo F, Wykowska A. Examining joint attention with the use of humanoid robots-A new approach to study fundamental mechanisms of social cognition. Psychonomic Bulletin and Review, 2020 Apr, 27(2), 217–236. doi: 10.3758/s13423-019-01689-4.

[40] Marino F, Chilà P, Sfrazzetto ST, et al. Outcomes of a robot-assisted social-emotional understanding intervention for young children with autism spectrum disorders. Journal of Autism and Developmental Disorders, 2020 Jun, 50(6), 1973–1987. doi: 10.1007/s10803-019-03953-x.

[41] Van Straten CL, Smeekens I, Barakova E, et al. Effects of robots' intonation and bodily appearance on robot-mediated communicative treatment outcomes for children with autism spectrum disorder. Personal and Ubiquitous Computing, 2018, 22, 379–390. https://doi.org/10.1007/s00779-017-1060-y

[42] Der Pütten AMR-V, Krämer NC, Herrmann J. The effects of humanlike and robot-specific affective nonverbal behavior on perception, emotion, and behavior. International Journal of Social Robotics, 2018, 10, 569–582.

[43] Jain S, Thiagarajan B, Shi Z, Clabaugh CE, Matari'c MJ. Modeling engagement in long-term, in-home socially assistive robot interventions for children with autism spectrum disorders. Science Robotics, 2020, 5, Assessed: March 14, 2014, at https://www.semanticscholar.org/paper/Modeling-engagement-in-long-term%2C-in-home-socially-Jain-Thiagarajan/5bd1533df3354f19696c701b946c1960b9a05cbd.)

[44] Richardson K. Challenging Sociality. In: An anthropology of robots, autism, and attachment. Berlin/Heidelberg, Germany: Springer International Publishing; 2018.

[45] Napoli CD, Rossi S. A layered architecture for socially assistive robotics as a service. 2019 IEEE International Conference on Systems, Man and Cybernetics (SMC); doi: 10.1109/smc.2019.8914532.

[46] Egido-García V, Estévez D, Corrales-Paredes A, Terrón-López MJ, Velasco-Quintana PJ. Integration of a social robot in a pedagogical and logopedic intervention with children: A case study. Sensors (Basel), 2020 Nov 13, 20(22), 6483. doi: 10.3390/s20226483.

[47] Kumazaki H, Yoshikawa Y, Muramatsu T, et al. Group-based online job interview training program using virtual robot for individuals with autism spectrum disorders. Frontiers in Psychiatry, 2022 Jan 24, 12, 704564. doi: 10.3389/fpsyt.2021.704564.

[48] Yoshikawa Y, Kumazaki H, Kato TA. Future perspectives of robot psychiatry: Can communication robots assist psychiatric evaluation in the COVID-19 pandemic era?. Current Opinion in Psychiatry, 2021 May 1, 34(3), 277–286. doi: 10.1097/YCO.0000000000000692.

[49] Guemghar I, Pires de Oliveira Padilha P, Abdel-Baki A, Jutras-Aswad D, Paquette J, Pomey MP. Social robot interventions in mental health care and their outcomes, barriers, and facilitators: Scoping review. JMIR Mental Health, 2022 Apr 19, 9(4), e36094. doi: 10.2196/36094.

[50] Nikopoulou VA, Holeva V, Tatsiopoulou P, Kaburlasos VG, Evangeliou AE. A pediatric patient with autism spectrum disorder and comorbid compulsive behaviors treated with robot-assisted relaxation: A case report. Cureus, 2022 Feb 20, 14(2), e22409. doi: 10.7759/cureus.22409.

[51]   Javed H, Park CH. Promoting social engagement with a multi-role dancing robot for in-home autism care. Frontiers in Robotics and AI, 2022 Jun 20, 9, 880691. doi: 10.3389/frobt.2022.880691.

[52]   Chevalier P, Ghiglino D, Floris F, Priolo T, Wykowska A. Visual and hearing sensitivity affect robot-based training for children diagnosed with autism spectrum disorder. Frontiers in Robotics and AI, 2022 Jan 12, 8, 748853. doi: 10.3389/frobt.2021.748853.

[53]   Kostrubiec V, Kruck J. Collaborative research project: Developing and testing a robot-assisted intervention for children with autism. Frontiers in Robotics and AI, 2020 Mar 31, 7, 37. doi: 10.3389/frobt.2020.00037.

Juan-Francisco Álvarez-Herrero

Proposal for intervention with intelligent educational robots in primary education students after the analysis of their pedagogical possibilities

Abstract: Intelligent educational robots (IERs) offer a variety of pedagogical possibilities that can enhance the development of critical thinking in primary education students. By having to solve challenges, experiment, analyze results, and continually improve, students not only acquire technical skills but also develop cognitive skills critical for success in learning. While IER offer numerous opportunities for innovative learning and skill development, they also pose challenges that must be addressed effectively. Based on the analysis of the existing literature on proposals with IER, the possibilities of these to improve learning are described, and a proposal for activities is made that takes into account the potential of these robots and also the challenges posed by their use, managing to create meaningful and transformative educational experiences for students. However, it is critical to address the challenges associated with these initiatives, such as teacher training, curricular alignment, and ethical considerations, to ensure their long-term effectiveness and success. Through this exhaustive and critical analysis, and a coherent, effective, and ethically responsible proposal of activities, we seek to contribute to the growth and evolution of this interdisciplinary field, with the ultimate objective of improving the quality and impact of primary education in the world.

Keywords: Educational robots, critical thinking, didactic proposal, primary education

1 Introduction

Today's society demands citizens trained in the most diverse disciplines. Current and future professions require that people be digitally competent and that they also have developed skills such as learning to learn, critical thinking, group work or collaborative work, and logical-mathematical thinking, in addition to the so-called soft skills [1–3].

Focusing attention on educational robotics, computational thinking, and programming, it is interesting to see how in the minimum educational contents of many countries, it is already requested to work on issues such as robotics and computational thinking from an early age [4, 5]. So, it is not surprising that educational robots and

---

Corresponding author: Juan-Francisco Álvarez-Herrero, Department of General and Specific Didactics, Faculty of Education, University of Alicante, Alicante, Spain, e-mail: juanfran.alvarez@ua.es

https://doi.org/10.1515/9783111352695-008

activities focused on bringing computational thinking and programming closer are increasingly present in classrooms.

But when we talk about educational robots, it is convenient to differentiate the different types of robots that exist and the uses that are made of them in education. A good approach is to resort to a taxonomy of these robots [6], and thus, in addition to verifying that there are more types of robots, being able to decide which ones are most appropriate for certain ages. In the case of early childhood education and the first years of primary education, in addition to floor robots [7], we have intelligent educational robots (IERs).

Therefore, the objective of this research is, after analyzing and verifying the possibilities of these resources, to propose a series of educational initiatives so that, with the use of IERs, students in the first years of primary education can develop critical thinking skills.

Despite the growing interest in the use of IERs in primary education, there are still challenges and questions that require further investigation. For example, how to design effective activities that make the most of the potential of robots for meaningful learning or how to guarantee accessibility and equity in access to this technology in different educational environments. Hence, this research tries to provide answers to these questions, proposing a series of activities to achieve better use of IERs in classrooms.

This chapter offers an in-depth exploration of the integration and impact of IERs in primary education. It begins with an introduction highlighting society's growing need to develop digital competences and critical thinking skills among tomorrow's citizens. It then delves into the characteristics and educational benefits of IERs, emphasizing their role in fostering cognitive and social skills through interactive and hands-on learning experiences. It then offers an educational proposal consisting of six varied activities that demonstrate how IERs can be effectively used to develop students' problem-solving skills, creativity, and teamwork. Each activity is designed with clear objectives, the materials needed to carry them out, and the assessment criteria needed to ensure their practical application in the classroom. The chapter also addresses the challenges and ethical considerations associated with implementing IERs in education, highlighting the importance of teacher training and equitable access to technology. Finally, by reading it, teachers and educational policy makers will gain valuable insights into the potential of IERs to improve learning outcomes and prepare students for future challenges in an increasingly technological world.

2 Intelligent educational robots (IERs) in education

IERs are revolutionizing the educational landscape, providing dynamic and interactive tools that enhance the learning experiences of students of all ages. In primary education, IERs are particularly effective in developing critical thinking and problem-solving skills. Furthermore, being equipped with advanced artificial intelligence (AI) and sensory capabilities, these robots are able to create engaging learning environments for students through hands-on experimentation and collaborative activities. Recent studies highlight the positive impact of IERs on cognitive and academic skills, emphasizing their role in promoting creativity, teamwork, and a deeper understanding of STEAM subjects.

2.1 Intelligent educational robots (IERs)

The IERs are increasingly present in our society. Little by little, they have been gaining a place among us, and this also means their implementation in education. From the earliest ages, with early childhood education students [8], to higher education students [9, 10], these robots are excellent resources that help in the teaching-learning processes. These robots, equipped with advanced AI capabilities and interactive design, offer a variety of opportunities to enrich the learning process and promote the development of critical thinking in early childhood students. From problem-solving to hands-on experimentation, IERs provide a dynamic and stimulating learning environment that encourages exploration, collaboration, and creativity.

Recent research has highlighted the potential positive impact of educational robots on the development of cognitive and academic skills in primary school students. For example, according to Ouyang and Xu [11], integrating educational robotics into the classroom can promote critical thinking and problem-solving through hands-on, project-based activities. Furthermore, studies such as that of Hong [12] and other researchers [13–15] have shown that the use of robots in the context of STEAM (science, technology, engineering, arts, and mathematics) education can significantly improve the understanding of scientific concepts and mathematics as well as encourage creative thinking and collaboration among students.

Some of the most common characteristics that IERs present are [16–19]:

- Accessible programming: It allows students to learn to program in an intuitive, sequenced, and progressive way, often using graphical interfaces or simplified programming languages (known as programming code).
- Social interaction: These robots can communicate with students in a natural and simple way, answering their questions or providing feedback. In this way, they are guided, and more personalized learning is occurring.
- Flexibility and adaptability: IERs can adapt to a wide variety of learning situations and skill levels, making them useful from primary education to higher education.
- Immediate feedback: They offer immediate feedback on student performance, allowing for more effective learning and quick correction of errors. Learning from mistakes involves effective and lasting learning.
- Advanced sensory capabilities: These robots typically come equipped with a wide variety of sensors (such as cameras, microphones, and proximity sensors) that

allow them to perceive what is happening in their environment and respond appropriately. The interaction offered in this way is very interesting for learning, especially for students at lower levels.

–   Full integration with the educational curriculum: They are designed to complement existing curricula and help educators teach difficult concepts more effectively, whether in STEM content or even other less scientific or technological subjects.

2.2 The IER and the development of critical thinking

But without a doubt, one of their main qualities is the excellent opportunity they have to develop critical thinking among students, especially those in the first years of primary education.

IERs have emerged as a promising tool in the educational field, especially in the development of critical thinking in primary education students. Critical thinking, defined as the ability to analyze, evaluate, and synthesize information in a reflective and creative manner, is a fundamental skill in learning and problem-solving in everyday life. In this context, educational robots offer a series of pedagogical possibilities that can enhance the development of critical thinking in students effectively [20].

Some of these possibilities are achieved, thanks to the fact that IERs are very good at working with them:

–   Problem-solving: Students can face specific challenges and complex problems that require them to use logic and reasoning to program the robot effectively and overcome obstacles by providing innovative solutions [21, 22]. When programming and controlling robot behavior, students must use critical thinking to identify problems, hypothesize, and develop effective strategies to overcome obstacles.

–   Experimentation and exploration: Educational robots offer an interactive learning environment where students can experiment with different strategies and observe how they affect the robot's behavior. This encourages active exploration and curiosity, key elements in the development of critical thinking [23].

–   Analysis of results: After programming a robot to perform a specific task, students can analyze the results obtained and reflect on what worked well and what could be improved. This helps them develop critical analysis and evaluation skills [24, 25].

–   Logical and sequential thinking: Bers and Elkind [26] point out that robot programming educates students in logical and sequential reasoning. When designing scripts and algorithms to guide the robot's behavior, students must think logically and analytically, considering the causes and effects of their actions.

–   Iteration and continuous improvement: As students experiment with robot programming, they have the opportunity to iterate on their designs and improve their solutions based on the feedback received [22, 27, 28]. This fosters the ability to critically reflect on their work and look for ways to improve it.

– Stimulating creativity: Liu et al. [29] highlight that educational robots offer a space for creativity and experimentation. Students can design and customize the appearance and behavior of robots, allowing them to explore innovative solutions and express their creativity in the learning process [23].

– Facilitation of collaborative work: Some authors [30–32] highlight that the use of educational robots promotes teamwork and collaboration among students. By working together to program and control the robots, students must communicate, negotiate, and share ideas, strengthening social and emotional skills key to developing critical thinking.

2.3 Advantages and disadvantages of using IER in classrooms

The use of IER in classrooms, and more specifically with primary education students for the development of students' scientific thinking, presents several advantages and disadvantages [33–36]:

Advantages:
– Development of digital competence: The use of these IER encourages the development of students' digital competence. In the management of these robots, skills related to digital technology, programming, and the management of technological devices, etc. are developed.

– Practical experience: Educational robots offer a unique opportunity for students to interact with scientific concepts in a practical and tangible way, which facilitates the understanding of abstract concepts that, in a traditional class, are quite inaccessible to primary education students.

– Stimulating interest and motivation in learning: Robot technology can spark students' interest and curiosity in learning in general, and science learning in particular, by providing dynamic and fun learning experiences. The incentive of handling and interacting with a technological device such as the IER provides satisfaction to the student, which conditions them to want to continue learning in that way.

– Safe experimentation: Educational robots allow students to conduct experiments and tests in a safe and controlled environment, reducing the risks associated with certain scientific activities.

– Technical skill development: By working with robots, students develop technical skills such as programming, problem-solving, and understanding technological concepts, which are essential in the field of science. At the same time, they can also practice oral expression and comprehension, which allows for the development of linguistic competence.

–   Personalization of learning: Educational robots can adapt to different skill levels and learning styles, allowing for more personalized and effective teaching. As already mentioned, the interaction and feedback that occur between the robot and the student allow the former to adapt, select, and propose activities that enable progress in the learning already consolidated or reinforce those that have not been fully achieved.

Disadvantages:
Although IERs offer numerous advantages for the learning of primary education students, they also pose a series of challenges and possible drawbacks that must be considered [37–40]. It is essential to critically address these aspects to ensure effective and equitable implementation of technology in the classroom:

–   Cost and accessibility: Educational robots can be expensive to purchase, which may limit their availability in some classrooms and schools. The price of these devices is not always affordable for the budgets of educational centers, making it difficult for them to be available in classrooms. This can create disparities in access to technology between different groups of students, exacerbating existing educational inequities and increasing the already existing digital divide.

–   Training need: Educators may require additional training to effectively use educational robots in the classroom, which may represent a barrier to implementation. The training of teachers, both those who are already active and those future teachers in training, is very necessary; unfortunately, it has to compete with other training needs to which the teacher is obliged. The latter is an extra aggravating factor and makes this one of the main problems of the implementation of IER in classrooms, or perhaps, the one that affects the most. As Negrini [41] comments, many teachers may lack the necessary preparation to use this technology effectively, which can limit its implementation and use in the educational process.

–   Possible technological dependency: There is a risk that excessive use of IERs could create technological dependency in students, decreasing their ability to solve problems independently. It is important to balance the use of technology with other teaching strategies that encourage critical thinking and problem-solving without relying exclusively on technological devices.

–   Limitations in conceptual learning: Although educational robots can be useful for teaching technical and procedural skills, it is important to ensure that the underlying scientific concepts are also addressed in a meaningful way. According to Xia and Zhong [42], this may require a more careful and deliberate approach in the design of learning activities. The same can be said when we talk about humanistic concepts and content since on many occasions the main objectives of education are ignored, which would be the development of skills that allow compliance with the evaluation criteria, as this demonstrates the learning of the students.

–   Curriculum integration challenges: Effectively integrating educational robots into the school curriculum can be challenging, as it requires coordination and collaboration between educators from different disciplines. As Bers and Elkind [26] highlight: "Coordination and collaboration between educators from different disciplines may be necessary to design activities that meaningfully integrate technology into student learning." Even when it is a single generalist teacher who works in all or almost all disciplines, there may be the mistake of considering educational robotics as something recreational or something to work on sporadically as a reward for their students, without getting to work on all the development of skills behind it.

– Difficulties associated with the operation and interaction of the IER with students: There is a difficulty associated with the fact that certain actions and interactions with the robot are sometimes, and more specifically at certain ages, difficult to understand, or there is a disconnection between what the IER aims to achieve and what the student believes he understands the robot is asking of him. This often occurs when the robot sends messages that are difficult to understand or that cause ambiguity and confusion for the student.

To all these disadvantages and returning to the topic that this research addresses, it must be said that since these are very recent resources (the IER), there is very little research in this regard as well as very few proposals and practices carried out with them.

3 Methodology

The methodology carried out consisted of first conducting an analysis of the bibliography to determine what types of proposals and activities would allow the use of the IER with primary education students to develop critical thinking. The search for sources and references led us to identify some of the proposals and activities carried out and, on the other hand, to propose the different types of activities that can occur with the IER. Finally, we made our own proposal of activities.

3.1 Analysis of proposals

IERs have been the subject of numerous educational proposals in recent years, which seek to take advantage of their potential to improve student learning and engagement in educational environments. Through an initial review of the existing literature, several effective activities and practices that have emerged in this context can be identified.

One of the focus areas is the integration of robots into the school curriculum, as suggested by Scaradozzi et al. [43]. These authors highlight the importance of designing activities that are aligned with the established learning objectives and the evaluation criteria to be achieved, which guarantees an effective integration of technology in the educational process. For example, Bers et al. [44] propose the use of robots in activities that combine STEM, thus promoting a holistic approach to learning.

On the other hand, Anwar et al. [36] carried out a systematic review on the usefulness of robots in the classroom. They highlighted the importance of designing activities that are aligned with learning objectives and that make the most of the robots' capabilities to improve student engagement and learning.

Another key aspect is the role that the teacher plays in the implementation of these activities. Usselman et al. [45] explored teachers' pedagogical decisions when implementing robotics kits in the primary and secondary education curriculum. These authors highlight the importance of teacher training and institutional support to guarantee the success of these initiatives. In Spain, the new educational law contemplates the implementation of educational robotics and computational thinking from the early childhood education stage.

In addition, the impact of activities with robots on the development of specific skills in students has been investigated. For example, Zhang and Zhu [21] examined how the use of robots can promote creativity skills in students. Their study suggests that interaction with robots can stimulate imagination and exploration, leading to positive results in the development of creativity.

However, there remain challenges and ethical considerations associated with the use of robots in the classroom that must continue to be analyzed and resolved. Serholt et al. [46] explore the ethical tensions perceived by teachers when implementing robots in the classroom, highlighting the importance of addressing these concerns to ensure ethical and equitable educational practices. The incorporation of AI into educational robots makes the ethical problems raised by these devices greater and therefore must be addressed before launching any pedagogical action with these robots.

The active participation of the teacher is essential for the success of these initiatives. Ntemngwa and Oliver [47] explore teachers' pedagogical decisions when implementing robotics kits in the primary and secondary education curriculum. These authors highlight the importance of teacher training and institutional support to ensure the effective implementation of technology in the classroom.

3.2 Types of activities that can be considered with the IER

As has been seen, the potential of IER to transform primary education is undeniable. Its ability to foster creativity, collaboration, and problem-solving opens new doors for experiential learning and active participation of students in their own educational process. It is the duty of teachers to take advantage of this opportunity to create stimulating and enriching learning environments that prepare students to face the challenges of the twenty-first century and become competent and committed citizens in an increasingly technological and globalized society.

To achieve this, below are some possible activities that can be generated to develop these capabilities:

-   Ask open questions. With the help and interaction of the IER, the student will be able to ask questions that require answers that go beyond a simple "yes" or "no." In this way, the expression of opinions and the justification of answers by the student are also encouraged.
-   Stimulation of curiosity: Thanks to the attractiveness of these robots with all their interaction possibilities and different activities, the student's curiosity is awakened. There, the participation of the teacher will be necessary, who will play the role of guide and advisor and will lead the student to ask questions such as "Why do you think this happens?" or "What would happen if . . .?" These, in turn, will lead the student to promote exploration and inquiry, fostering a taste for research.
-   Problem-solving approach: The teacher, with the help of the IER, will pose challenges that require the student to find solutions. Thus, activities such as puzzles, construction games, and mathematical problems that can be posed with the help of the robot will be excellent mechanisms that allow students to develop problem-solving and logical reasoning skills.
-   Conducting guided debates between the teacher, the robot, and the student. Group discussions can be held on topics relevant to the level of understanding of the students, in which the participation of the robot is relevant. This will encourage students to express their opinions and also know how to listen to those of others, thus promoting critical thinking and empathy. At these ages, making students empathetic is very important for their personal development.
-   Critical literature activities: From the stories and narratives that the robot shares, ethical dilemmas or problematic situations can be raised, in which, with the help and participation of the teacher, the interaction can be followed, creating situations in which the student has to ask themselves what they would do in a situation similar to the one presented in the story, and why they would do it that way and not another.
-   Classification and categorization activities: With the help of the robot and the teacher, lists of objects, things, or animals can be provided, and the teacher can ask their students to try to classify said objects, animals, or things according to different criteria. With this type of practice, students are encouraged to organize information and recognize patterns.
-   Role-playing games: Based on the situations and interactions that may occur in the IER, the teacher will provoke situations in which students can assume different roles and thus find themselves in the position of having to make decisions. This will allow them to see the consequences of their choices and also develop decision-making skills.
-   Encouragement of observation: With the complicity of the robots and the role of the teacher as a guide, students will be asked to observe their environment and

ask questions about what they see around them. This encourages students to develop critical observation skills as well as the ability to make connections between different concepts.

– Creative projects: Again, with the complicity of the robot and the supervision and guidance of the teacher, students will be asked to carry out different projects (within the project-based learning methodology), where they must plan, execute, and evaluate their own creations. This develops critical thinking skills throughout the entire process involved in completing these projects.

– Problem-based learning: Through this methodology, students, with the intervention of robots and teachers, will be able to address topics through practical situations and real-world problems, always from the perspective of a familiar and known environment. In this way, students are invited to investigate and find solutions, promoting applied critical thinking.

4 Proposal for activities

We live in an age where technology is seamlessly integrated into our daily lives, so the use of IER in the classroom represents a transformative approach to learning, as seen above. This proposal outlines a series of engaging activities designed to fully leverage the power of robotics to enhance students' logical thinking, creativity, and collaboration skills. Through six hands-on experiences such as navigating mazes, organizing robot parades, or solving secret codes, students will not only understand fundamental programming concepts but also develop essential competencies in several disciplines including math, science, and the arts. By fostering teamwork and problem-solving skills, these activities aim to create an interactive learning environment that prepares students for the challenges of the future while making education an incredibly engaging adventure for them.

4.1 Activity 1: inside the labyrinth

Description: Students will program an educational robot to navigate through a maze and be able to exit it.
Steps to follow:
Introduce basic programming concepts such as sequential instructions and loops.
Divide students into groups and provide them with an educational robot and a simple maze.
Students will program the robot to navigate through the maze using basic instructions such as "go forward" and "turn left."
Objectives: Develop logical and sequential thinking skills as well as problem-solving skills.

Skills that are developed: Mathematical and digital skills, and learning to learn.
Benefits: Encourages teamwork, problem-solving, and creativity.
Relationships with other areas: Mathematics (coordinates, directions), sciences (concepts of movement and energy).
Necessary material resources: Educational robots, printed or block-built mazes, and adequate space for the activity.
Evaluation: Observation of student performance during programming and problem-solving. Evaluation of the robot's accuracy and efficiency when navigating the maze.
Timing: 2 sessions of 45 min.

4.2 Activity 2: robotic parade

Description: Students will design and program robots to carry out a thematic parade.
Steps to follow:
Introduce basic design and programming concepts.
Students will work in groups to design and decorate their robots according to a specific theme.
They will program the robot's movements so that it "parades" in a synchronized manner.
Objectives: Promote creativity, collaboration, and artistic expression through technology.
Competencies that are developed: Artistic and cultural competence, digital competence, and social and civic competence.
Benefits: Promotes collaboration, creativity, and self-expression.
Relationships with other areas: Artistic education (design and decoration), physical education (coordination of movements).
Necessary material resources: Educational robots, decoration materials, and music for the parade.
Evaluation: Evaluation of the originality of the design, the coordination of the robot's movements, and team collaboration.
Timing: 3 sessions of 45 min.

4.3 Activity 3: secret code

Description: Students will use codes and clues to program a robot and decipher a secret message.
Steps to follow:
Introduce the basic concepts of programming and logic.
Provide students with a series of codes and clues to decipher a message.

Students will program the robot to follow instructions and pick up clues hidden around the classroom.

Objectives: Develop logical thinking, problem-solving, and collaboration skills.

Competencies that are developed: Linguistic competence, digital competence, learning to learn.

Benefits: Encourages teamwork, problem-solving, and critical thinking.

Relationships with other areas: Language (reading comprehension and writing messages), mathematics (patterns and sequences).

Necessary material resources: Educational robots, printed clues, encrypted messages.

Evaluation: Evaluation of the time needed to decipher the message, the precision in the robot's programming, and team collaboration.

Timing: 2 sessions of 45 min.

4.4 Activity 4: space explorers

Description: Students will use robots to explore and map a simulated "planet" in the classroom.

Steps to follow:

Introduce the basic concepts of navigation and cartography.

Students will work in groups to program the robot and map the "planet" terrain.

They will use sensors and programming to avoid obstacles and record the terrain explored.

Objectives: Develop cartography, navigation, and teamwork skills.

Competencies that are developed: Mathematical competence, digital competence, social and civic competence.

Benefits: Encourages collaboration, problem-solving, and spatial thinking.

Relationships with other areas: Geography (map and terrain concepts), sciences (spatial exploration).

Material resources needed: Educational robots, simulated obstacles, and large paper to create the "map" of the planet.

Evaluation: Evaluation of the accuracy of the created map, the efficiency of the robot's navigation, and team collaboration.

Timing: 3 sessions of 45 min.

4.5 Activity 5: robotic dance

Description: Students will program robots to perform a dance choreography (Fig. 1).

Steps to follow:

Introduce basic concepts of programming and the coordination of movements.

Students will work in groups to program the robot and create a dance choreography.

They will use music and synchronized movements to perform the programmed dance.

Objectives: Promote creativity, coordination, and collaboration.

Competencies that are developed: Artistic and cultural competence, digital competence, social and civic competence.

Benefits: Promotes artistic expression, team collaboration, and creative thinking.

Relationships with other areas: Artistic education (dance and body expression), Physical education (coordination of movements).

Necessary material resources: Educational robots, music, and adequate space for dancing.

Evaluation: Evaluation of the originality of the choreography, the precision in programming the robot's movements, and the synchronization with the music.

Timing: 3 sessions of 45 min.

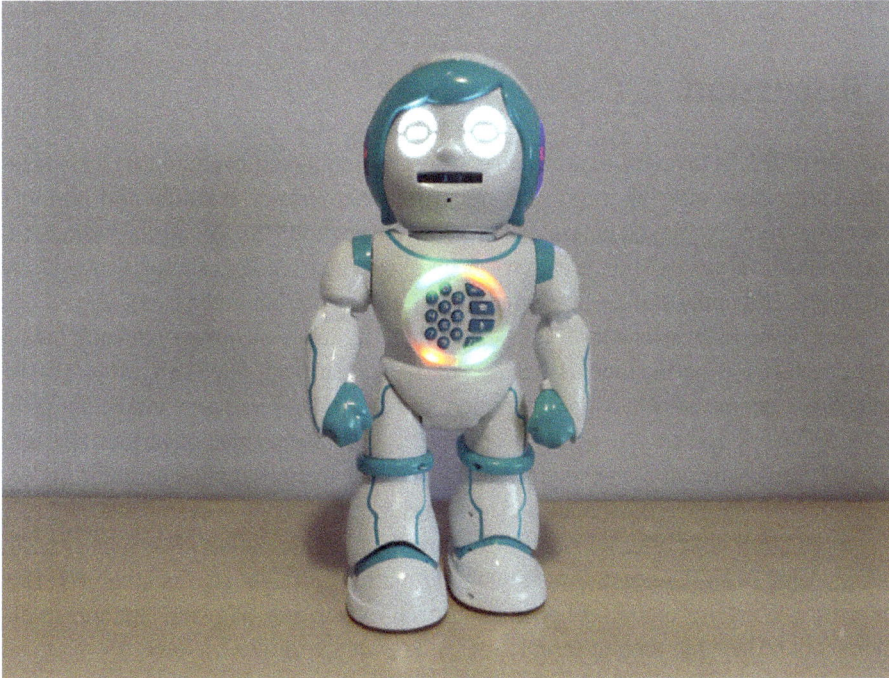

Fig. 1: An IER dancing with movements coordinated to the music.

4.6 Activity 6: obstacle course

Description: Students will program robots to overcome a series of obstacles in a race.

Steps to follow:

Introduce the basic concepts of programming and problem-solving.

Divide students into teams and provide them with an educational robot and an obstacle course.

Teams will program their robots to overcome obstacles and reach the finish line.

Objectives: Develop problem-solving, teamwork, and strategic thinking skills.

Competencies that are developed: Mathematical competence, digital competence, social and civic competence.

Benefits: Encourages collaboration, creativity, and programming proficiency.

Relationships with other areas: Physical education (coordination of movements), mathematics (measurement of distances and times).

Necessary material resources: Educational robots, obstacle courses, and stopwatches.

Evaluation: Evaluation of the robot's efficiency in overcoming obstacles, team collaboration, and the strategy used.

Timing: 2 sessions of 45 min.

5 Discussion

The use of IER in primary education represents a significant evolution in the educational landscape, offering innovative opportunities to enrich learning and promote the comprehensive development of students. Throughout this research, various aspects related to this exciting area have been explored, from educational proposals to challenges and opportunities presented by its implementation in the classroom.

One of the main strengths of IERs lies in their ability to encourage critical thinking and problem-solving in students from an early age. Through hands-on, project-based activities, students have the opportunity to face real challenges, design innovative solutions, and work as a team to achieve common goals. As highlighted in various research [36, 44], these active and meaningful learning experiences are fundamental for the development of cognitive, emotional, and social skills in primary school students.

In addition, IERs offer a versatile platform for the integration of various curricular concepts from science and technology to mathematics, languages, and arts. This interdisciplinarity promotes a holistic approach to learning, allowing students to explore connections between different areas of knowledge and understand the practical relevance of what they learn in the classroom [26, 43].

However, the use of IERs also poses challenges that must be addressed to ensure their effectiveness and equity in the classroom. Teacher training, technological accessibility, and ethical considerations are just some of the aspects that require careful attention from educators, administrators, and educational policy makers [46, 47].

6 Conclusions

May this analysis and didactic proposal presented here serve to open the way for the implementation of the IER in primary classrooms as a first step to continue working on educational robotics, computational thinking, and programming in subsequent educational stages. With this, we aim to encourage the development of skills such as critical thinking, learning to learn, collaborative work, problem-solving, creativity, and logical-mathematical thinking.

The use of IER in primary education therefore represents a bold step toward the future of teaching and learning. With careful planning, effective collaboration, and a continued commitment to educational innovation, the potential of this technology can be fully harnessed to create meaningful and transformative educational experiences for all students. A potential that, as has been seen, is significant and promising. All this must be done without losing sight of providing quality training in the use of these robots to both teachers in training and active teachers as well as providing the necessary resources to educational centers and not neglecting the ethical considerations that its use implies.

From all the proposals raised, it is clear that, although the IER are excellent resources to promote and develop critical thinking, as well as many other skills in students, all of this would not be possible without mediation and the role of counselor and guide that the teacher must have. If the teacher did not intervene in all these interaction proposals, there would be no learning and everything would remain a simple game.

Therefore, it is necessary to give visibility to the magnificent opportunities that IERs present for the development of student competencies, but at the same time, it must also be accompanied by teacher training plans so that they can guide this type of practice and enable the learning. A learning that will also be authentic and lasting, since the student has had to interact with the robot and the teacher, but without at any time losing the prominence that this type of learning and these resources give him.

Integrating IERs into educational systems is a promising tool to transform teaching methodologies and student engagement. By fostering an interactive and personalized learning environment, IERs not only enhance cognitive and academic skills but also enable the development of essential twenty-first-century competencies such as collaboration, critical thinking, and creativity. Future research that is intended to be conducted includes exploring the long-term impacts of IERs on educational outcomes, exploring their potential in special education, and researching and developing more sophisticated adaptive learning systems powered by AI and IERs. Furthermore, research into the ethical and accessibility implications of IERs will also be a major focus of study, as this will be crucial to ensuring equitable benefits across diverse educational settings.

References

[1] Marín Suelves D, Gabarda Méndez V, Ramón-Llin Mas JA. Análisis de la competencia digital en el futuro profesorado a través de un diseño mixto. Red [Internet], 2022, 22(70), https://revistas.um.es/red/article/view/523071

[2] Dondi M, Klier J, Panier F, Schubert J. Defining the skills citizens will need in the future world of work. McKinsey & Company, 2021, 25, 1–19.

[3] Rakowska A, de Juana-Espinosa S. Ready for the future? Employability skills and competencies in the twenty-first century: The view of international experts. Human Systems Management, 2021, 40(5), 669–684.

[4] Valls Pou A, Canaleta X, Fonseca D. Computational Thinking and Educational Robotics Integrated into Project-Based Learning. Sensors, 2022, 22(10), 3746. https://doi.org/10.3390/s22103746

[5] Piedade J, Dorotea N, Pedro A, Matos JF. On Teaching Programming Fundamentals and Computational Thinking with Educational Robotics: A Didactic Experience with Pre-Service Teachers. Education Sciences, 2020, 10(9), 214. https://doi.org/10.3390/educsci10090214

[6] Álvarez-Herrero JF. Diseño y validación de un instrumento para la taxonomía de los robots de suelo en Educación Infantil: [Design and validation of an instrument for the taxonomy of floor robots in Early Childhood Education]. Pixel-Bit, 2021, 60, 59–76. https://doi.org/10.12795/pixelbit.78475

[7] Álvarez-Herrero JF, Martinez-Roig R, Urrea-Solano M. Taxonomy of Floor Robots for Working on Educational Robotics and Computational Thinking in Early Childhood Education from a STEM Perspective. In: Papadakis S, Kalogiannakis M, eds. STEM, Robotics, Mobile Apps in Early Childhood and Primary Education. Lecture Notes in Educational Technology. Singapore: Springer; 2022. pp. 1–12. https://doi.org/10.1007/978-981-19-0568-1_12

[8] Jin L. Investigation on potential application of artificial intelligence in preschool children's education. Journal of Physics: Conference Series, 2019, 1288(1), 1–5.

[9] Al-Billeh T. Teaching law subjects by using educational robots: Does the use of robots lead to the development of legal skills among law staudents?. Asian Journal of Legal Education, 2024, 0(0), https://doi.org/10.1177/23220058241227610

[10] Zheng Y, Meng H, Jia W. Application Research and Challenges of Artificial Intelligence in Primary and Secondary Education. In: 2022 12th International Conference on Information Technology in Medicine and Education (ITME). Xiamen, China; 2022. pp. 162–166. https://doi.org/10.1109/ITME56794.2022.00043

[11] Ouyang F, Xu W. The effects of educational robotics in STEM education: A multilevel meta-analysis. International Journal of STEM Education, 2024, 11, 7. https://doi.org/10.1186/s40594-024-00469-4

[12] Hong KC. STEM/STEAM Education with Robotics in Elementary School Curriculum. International Conference on Convergence Technology, 2013, 2(1), 8–9.

[13] Kalaitzidou M, Pachidis TP. Recent Robots in STEAM Education. Education Sciences, 2023, 13(3), 272. https://doi.org/10.3390/educsci13030272

[14] Kim JO, Kim J. Development and Application of Art Based STEAM Education Program Using Educational Robot. In: Robotic Systems: Concepts, Methodologies, Tools, and Applications, edited by Information Resources Management Association. Hershey, PA: IGI Global; 2020. pp. 1675–1687. https://doi.org/10.4018/978-1-7998-1754-3.ch080

[15] Barnes J, FakhrHosseini MS, Vasey E, Duford Z, Jeon M. Robot theater with children for STEAM education. Proceedings of the Human Factors and Ergonomics Society Annual Meeting, 2017, 61(1), 875–879. Sage CA: Los Angeles, CA: SAGE Publications

[16] Bellas F, Sousa A. Computational intelligence advances in educational robotics. Frontiers in Robotics and AI, 2023, 10, 1150409.

[17] Eguchi A. Educational robotics for promoting 21st century skills. Journal of Automation, Mobile Robotics and Intelligent Systems, 2014, 8(1), 5–11.

[18] Chatzichristofis SA. Recent advances in educational robotics. Electronics, 2023, 12(4), 925. https://doi.org/10.3390/electronics12040925

[19] Khairy D, Abougalala RA, Areed MF, Atawy SM, Alkhalaf S, Amasha MA. Educational robotics based on artificial intelligence and context-awareness technology: A framework. Journal of Theoretical and Applied Information Technology, 2020, 98(13), 2227–2239.

[20] Zhang Y, Luo R, Zhu Y, Yin Y. Educational robots improve K-12 students' computational thinking and STEM attitudes: Systematic review. Journal of Educational Computing Research, 2021, 59(7), 1450–1481. https://doi.org/10.1177/0735633121994070

[21] Zhang Y, Zhu Y. Effects of educational robotics on the creativity and problem-solving skills of K-12 students: A meta-analysis. Educational Studies, 2022, https://doi.org/10.1080/03055698.2022.2107873

[22] Chevalier M, Giang C, Piatti A, Mondada F. Fostering computational thinking through educational robotics: A model for creative computational problem solving. International Journal of STEM Education, 2020, 7, 39. https://doi.org/10.1186/s40594-020-00238-z

[23] Hou H, Zhang X, Wang D. Can educational robots improve student creativity: A meta-analysis based on 48 experimental and quasi-experimental studies. Journal of East China Normal University (Educational Sciences), 2022, 40(3), 99–111. https://doi.org/10.16382/j.cnki.1000-5560.2022.03.009

[24] Ronsivalle GB, Boldi A, Gusella V, Inama C, Carta S. How to Implement Educational Robotics' Programs in Italian Schools: A Brief Guideline According to an Instructional Design Point of View. Technology, Knowledge and Learning, 2019, 24, 227–245. https://doi.org/10.1007/s10758-018-9389-5

[25] Souza ML, Andrade WL, Sampaio MR. Analyzing the Effect of Computational Thinking on Mathematics through Educational Robotics. In: 2019 IEEE frontiers in education conference (FIE). Covington, KY, USA; 2019. pp. 1–7. https://doi.org/10.1109/FIE43999.2019.9028384

[26] Bers MU, Elkind D. Blocks to robots: Learning with technology in the early childhood classroom. New York, USA: Teachers College Press; 2008.

[27] Catlin D, Woollard J. Educational robots and computational thinking. Proceedings of 4th International Workshop Teaching Robotics, Teaching with Robotics & 5th International Conference Robotics in Education, 2014, 1, 144–151.

[28] Paucar-Curasma R, Villalba-Condori K, Arias-Chavez D, Nguyen-Thinh Le, Garcia-Tejada G, Frango-Silveira I. Evaluation of computational thinking using four educational robots with primary school students in Peru. Education in the Knowledge Society, 2022, 23, https://doi.org/10.14201/eks.26161

[29] Liu X, Gu J, Zhao L. Promoting primary school students' creativity via reverse engineering pedagogy in robotics education. Thinking Skills and Creativity, 2023, 49, 101339. https://doi.org/10.1016/j.tsc.2023.101339

[30] Smakman M, Vogt P, Konijn EA. Moral considerations on social robots in education: A multi-stakeholder perspective. Computers & Education, 2021, 174, 104317. https://doi.org/10.1016/j.compedu.2021.104317

[31] Smakman M, Jansen B, Leunen J, Konijn E. Acceptable Social Robots in Education: A Value Sensitive Parent Perspective. In: INTED2020 Proceedings. IATED; 2020. pp. 7946–7953. https://doi.org/10.21125/inted.2020.2161

[32] Demetroulis FA, Theodoropoulos A, Wallace M, Poulopoulos V, Antoniou A. Collaboration skills in educational robotics: A methodological approach – results from two case studies in primary schools. Education Sciences, 2023, 13(5), 468. https://doi.org/10.3390/educsci13050468

[33] Tzagkaraki E, Papadakis S, Kalogiannakis M. Exploring the Use of Educational Robotics in Primary School and Its Possible Place in the Curricula. In: Educational robotics international conference. Cham: Springer International Publishing; 2021, pp. 216–229. https://doi.org/10.1007/978-3-030-77022-8_19

[34] Pozzi M, Prattichizzo D, Malvezzi M. Accessible educational resources for teaching and learning robotics. Robotics, 2021, 10(1), 38. https://doi.org/10.3390/robotics10010038

[35] Kyriazopoulos I, Koutromanos G, Voudouri A, Galani A. Educational Robotics in Primary Education: A Systematic Literature Review. In: I. management association, ed. research anthology on computational thinking, programming, and robotics in the classroom. Hersey, PA, USA: IGI Global; 2022. pp. 782–806. https://doi.org/10.4018/978-1-6684-2411-7.ch034

[36] Anwar S, Bascou NA, Menekse M, Kardgar A. A systematic review of studies on educational robotics. Journal of Pre-College Engineering Education Research (J-PEER), 2019, 9(2), 2. https://doi.org/10.7771/2157-9288.1223

[37] Papadakis S, Vaiopoulou J, Sifaki E, Stamovlasis D, Kalogiannakis M. Attitudes towards the Use of Educational Robotics: Exploring pre-service and in-service early childhood teacher profiles. Education Sciences, 2021, 11(5), 204. https://doi.org/10.3390/educsci11050204

[38] Zviel-Girshin R, Kukliansky I, Rosenberg N. Twenty-first century parents' attitudes and beliefs on early childhood robotics education. Education and Information Technology, 2023, https://doi.org/10.1007/s10639-023-12218-1

[39] Garvis S, Keane T. A Literature Review of Educational Robotics and Early Childhood Education. In: Garvis S, Keane T, eds. Technological innovations in education: Applications in education and teaching. Singapore: Springer; 2023. pp. 71–83. https://doi.org/10.1007/978-981-99-2785-2_6

[40] Tzagaraki E, Papadakis S, Kalogiannakis M. Teachers' Attitudes on the Use of Educational Robotics in Primary School. In: Papadakis S, Kalogiannakis M, eds. STEM, robotics, mobile apps in early childhood and primary education. lecture notes in educational technology. Singapore: Springer; 2022. pp. 1–13. https://doi.org/10.1007/978-981-19-0568-1_13

[41] Negrini L. Teachers' attitudes towards educational robotics in compulsory school. Italian Journal of Educational Technology, 2020, 28(1), 77–90. https://doi.org/10.17471/2499-4324/1136

[42] Xia L, Zhong B. A systematic review on teaching and learning robotics content knowledge in K-12. Computers & Education, 2018, 127, 267–282. https://doi.org/10.1016/j.compedu.2018.09.007

[43] Scaradozzi D, Sorbi L, Pedale A, Valzano M, Vergine C. Teaching robotics at the primary school: An innovative approach. Procedia-Social and Behavioral Sciences, 2015, 174, 3838–3846.

[44] Bers M, Seddighin S, Sullivan A. Ready for Robotics:Bringing together the T and E of STEM in early childhood teacher education. Journal of Technology and Teacher Education, 2013, 21(3), 355–377. https://www.learntechlib.org/primary/p/41987/

[45] Usselman M, Ryan M, Rosen JH, Koval J, Grossman S, Newsome NA, Moreno MN. Robotics in the core science classroom: Benefits and challenges for curriculum development and implementation (rtp, strand 4). 2015 ASEE Annual Conference & Exposition, 2015, 12315, 1–16

[46] Serholt S, Barendregt W, Vasalou A, Alves-Oliveira P, Jones A, Petisca S, Paiva A. The case of classroom robots: Teachers' deliberations on the ethical tensions. Ai & Society, 2017, 32, 613–631. https://doi.org/10.1007/s00146-016-0667-2

[47] Ntemngwa C, Oliver JS. The implementation of integrated science technology, engineering and mathematics (STEM) instruction using robotics in the middle school science classroom. International Journal of Education in Mathematics, Science and Technology, 2018, 6(1), 12–40.

Pedro Baena-Luna and Esther García-Río

New realities in the education field? Keys to understanding the role of intelligent educational robots

Abstract: Using robots is recognized as an innovative tool for improving learning processes. This is why many of the works that have addressed these realities have highlighted how it is a new technique that could change current educational approaches and facilitate students' learning in different environments. This chapter aims to conduct a bibliometric analysis of the scientific literature that connects the relationship between the realities of intelligent educational robots in educational environments. The scientific databases consulted were Web of Science (WoS) and Scopus. From an initial number of 47 and 820 works, respectively, once the results were unified, 169 were finally selected for analysis. The information was processed using the Bibliometrix tool, which provided information on the annual production and analysis of journals, authors, documents, keywords, etc. The results will allow us to identify the principal research trends in this area, establish relationships between them, and detect future research opportunities.

Keywords: Intelligent educational robots, evolution of education, Bibliometrix

1 Introduction

The interdisciplinary nature of robotics has motivated increased interest in the educational community regarding the possibilities of use and application in this field in recent years [1]. Educational robots can be a relevant tool, thanks to their ability to promote innovation, help teachers and students develop problem-solving skills and all kinds of content, and improve knowledge in different curricular fields [2, 3]. This is not surprising if we start from the premise that students' educational context nowadays is characterized by the ease of access to information through different technologies [4].

The use of robots is therefore recognized as an innovative tool for the improvement of learning processes. This is why many of the works that have addressed these realities have highlighted how it is a new technique that could change current educational approaches and facilitate students' learning in different environments [5]. This

Corresponding author: Pedro Baena-Luna, Department of Business Management and Marketing, University of Seville, Seville, Spain, e-mail: pbaenaluna@us.es
Esther García-Río, Department of Economic Analysis and Political Economy, University of Seville, Seville, Spain, e-mail: egrio@us.es

https://doi.org/10.1515/9783111352695-009

is a straightforward consequence of the evident change in the characteristics of today's students, which makes it necessary to adopt new pedagogical approaches to learning in educational environments [6].

The continuous advances in information and communication technologies have led teachers and students to learn to coexist with machines, favoring collaboration to obtain and effectively achieve specific results in the educational environment [7]. Despite the different experiences implemented in using robots for educational purposes, focusing on more than just using technology as an end is essential. To this end, it must be part of a plan shared by all the actors involved and affected by the use of robots for educational purposes [8].

The use of robotics is a phenomenon that has been around for a while. Its beginnings date back to the 1980s in the twentieth century when the first models of robots specifically designed for educational activities appeared and were marketed by Lego in 1988 [9]. Despite the enormous research efforts to extract and show the improvements in learning processes as a consequence of using these tools, there are still some gaps and areas where knowledge about them and their impact on these processes can be deepened and improved [10].

Given the above, it is necessary to explore research works in these fields further to unveil the mechanisms that will favor the integration of intelligent educational robots in educational processes. A bibliometric analysis of the above literature may shed light on the different current and past approaches in academia's treatment of the two realities in a connected way. Consequently, the following research questions are formulated:

RQ1. Which authors are at the forefront of scientific production in the related scientific literature?

RQ2. Which works have had the highest impact in subsequent related works?

RQ3. What are the main topics investigated, which countries lead this scientific production, and which keywords are the most used in the scientific literature analyzed?

RQ4. What conceptual, intellectual, and social structures were generated from the scientific literature analyzed?

RQ5. What are the future applications of robotic technology in education?

This chapter will attempt to answer these questions. The main contributions are the results obtained from the deep analysis of these realities in a connected way: intelligent educational robots and educational environments. Another relevant fact is that it is one of the few works that, in its analysis, combines the results of two databases (Scopus and WoS) to perform a single integrated analysis.

After this introduction, the rest of the document is organized as follows: the literature review section examines the research considered for the study and the background of the concepts. The following section describes the methodology, and the fourth section contains the results. In the subsequent section, the findings are discussed. Finally, the last section presents the most relevant conclusions of the work, future research lines, and the study's limitations.

2 Literature review

The use of robots as a tool to promote learning in educational environments is an increasingly recognized innovative resource [11] at all educational levels [12]. Educational robots improve student learning and transform more traditional approaches through new and attractive learning options [13]. Educational robots can be used in developing classroom activities included in the official teaching program and other external activities of a more playful nature [14], generating different possible scenarios for learning [15].

Due to the actuality and the recent incorporation of these robots into educational contexts, the definitions and concepts related to their use are still varied and even depend in many cases on the type of robots to which they refer, e.g., articulated robots, mobile robots, or autonomous robots and vehicles [16]. In general, and according to Pivetti et al. [14], they can be defined as robots that can interact with students. This is because students can give (or generate) instructions to the robot, and the robot responds to those requests.

Nowadays, in the twenty-first century, education is not only characterized by teaching content and learning it independently and unconnectedly [17]. The current educational model requires and is based on the development of communication skills, critical thinking, collaboration, etc. [4], in addition to other competencies and skills that are essential for the further development of the person, requirements that can be favored in achieving their goals using educational robots [18].

The interdisciplinary nature of these robots, together with their ability to interact in a friendly way, has made them a potential problem-solving tool for students from different disciplines [1]. Previous works have highlighted some of the positive effects such as improving critical thinking, algorithmic thinking, and creativity, and favoring the group cohesion of students [19].

Along with the benefits of using these robots in different disciplines, integrating these technological resources in parallel learning and teaching processes supports and reinforces students' possible interest in subjects related to science and technology that they have not shown so far [11, 18]. This is a consequence of using these robots, which involve the development of tasks through programmable elements, sensors, and other eminently technological tools [10].

In addition to the benefits and potential of using robots in the educational context, it is also true that their use and implementation pose a series of short and medium-term challenges that must be addressed in any case. As pointed out by Martínez-Comesaña et al. [20], the implementation of robots in the educational context requires the qualification of the personnel in charge of handling these robots, in addition to the creation and programming of the educational content, sometimes resulting in the rejection of this educational resource by some teachers because of the lack of this type of technological competence [21]. Another aspect highlighted by Martínez-Comesaña et al. [20] is the difficulty in providing individualized attention due to the lack of adaptation of this tool to the individual needs of the students and their learning styles.

3 Methodology

Bibliometrics is a mature literature analysis and information extraction method that objectively evaluates scientific research, offering advantages in quantitative and modeled research [22]. For this reason, its use is spreading to many disciplines. It is particularly suitable for investigating voluminous, fragmented, and controversial research streams [23]. Bibliometric analysis is a popular and rigorous method for exploring and analyzing large volumes of scientific data [24].

The decision to adopt a bibliometric method is based on the fact that research studies with data are considered more relevant than subjective evaluations through traditional reviews [25]. This method allows one to analyze the evolution of a specific field while highlighting emerging areas in that field [24].

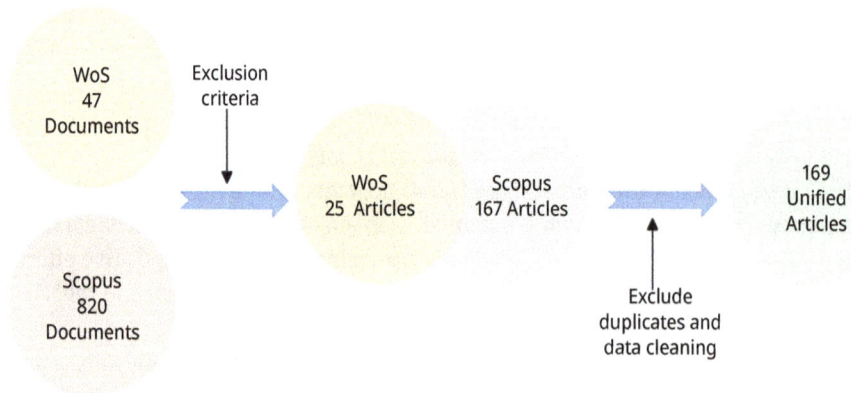

Fig. 1: Flowchart of research methodology.

The procedure consists of five steps based on the workflow recommended by Zupic et al. [26]. In the first step, the research questions were defined, and the categories of the analysis were selected together with the appropriate bibliometric indicators that could answer these questions. These indicators are structured into three types: performance indicators (measure the productivity of a given researcher), quality indicators (measure the quality of a researcher's production), and structural indicators (measure the connections between publications, authors, and research areas) [27].

Second, we selected the databases containing bibliometric data, filtered the primary documents by the chosen search criteria, and exported the data from the databases for unification. The selected databases were WoS and Scopus. The WoS database is essential for global scholarly information, and it has included more than 8,700 core academic journals in various fields since 1900 [28]. The choice of Scopus is due to its largest database of peer-reviewed literature abstracts and citations [29], with more than 41,000 source titles (journals, conference proceedings, book series, and trade publications) from approximately 12,500 publishers, of which approximately 94% are peer-reviewed [30]. The search performed in WoS Core Collection has been "intelligent robots" and "education" (topic) and in Scopus (TITLE-ABS-KEY ("intelligent robots" and "education")). The first search results returned 47 documents in WoS and 820 in Scopus. The exclusion criteria were applied, and only the article type was selected and published in English. Unifying the databases made it possible to eliminate duplicate documents and clean the dataset, resulting in 169 articles for bibliometrics.

Subsequently, the bibliometric analysis software tools were reviewed. The objective of this research is to combine the information collected in the WoS and Scopus databases, for which it was decided to use R Studio to unify the two sets of results and the Bibliometrix package with its Biblioshiny interface as software tools for bibliometric analysis. Before processing the data with this software, the following steps were taken: (1) access the web version of RStudio, (2) in the console window, type: install. packages("bibliometrix") and then library(bibliometrix), and (3) type the command to run the Biblioshiny program: biblioshiny [31]. Biblioshiny features a menu incorporating analysis and graphs for three-level metrics (source, author, and document) and three knowledge structures (conceptual, intellectual, and social) [32]. The process is summarized in Fig. 1.

Finally, the results have been graphically represented and interpreted. Only a few bibliometric studies merge the two databases to perform a single integrated analysis [33]. The methodology used in this bibliometric analysis provides the most relevant value by jointly analyzing the documents obtained in the WoS and Scopus databases.

4 Results

Bibliometric analysis techniques are manifested in two categories: (1) performance analysis and (2) scientific mapping. Performance analysis provides information on publications, while scientific mapping focuses on the relationships between research components. In performance analysis, the most prominent measures are the number of publications as an indicator of productivity and citations per year as a measure of impact; other measures, such as citations per publication or the h-index that combines citations and publications, are also used to measure performance. On the other hand, scientific mapping focuses on intellectual interactions and structural connections between research components (citation analysis, co-citation analysis, bibliographic linkage, co-word analysis, and coauthorship analysis) [24].

After data processing through Bibliometrix, the results show a graphical, statistical, and factorial analysis of the 169 documents studied. Metrics have been obtained for the study of performance: total publications, number of authors (single author and coauthorship), number of active years of publication, productivity per publication year, metrics related to accurate and average citations, and the collaboration index. This technique has been used to determine the trend or trajectory of intelligent educational robots and educational environments and map the intellectual structure. These results provide information on related research's past, present, and future. A summary of the data analyzed is shown in Tab. 1.

The first publication dates to 1996, but only at the beginning of the millennium did academic interest in the subject of study begin to awaken. Figure 2 shows the evolution of the number of annual publications. The trend line shows progressive growth since 2001, with a turning point in 2009. From this year until 2015, there was a downward trend, followed by a recovery of progressive growth until the present day. The scientific production followed a regression model with an increasing trend and a coefficient of determination $R^2 = 53.98\%$. The increase in the number of publications may be related to the birth of the term "educational robotics" as a subdiscipline of robotics applied to the educational field. Educational robotics focuses on robot design, analysis, application, and operation. Educational robots and robotics competitions have become prevalent educational activities that actively engage children in critical thinking and problem-solving in the classroom. Consequently, there has been an increase in the number of research studies in this field [34].

Figure 3 shows the total number of citations over the years. The average number of citations per year shows an increasing trend, with the beginning of growth coinciding with the year in which interest in the subject takes off and the number of documents published yearly begins to increase: 2018 and 2020, with 9.10 and 7.69 average citations per year, are the years with the highest average citations.

Table 2 shows the information regarding the research quality of the most prolific authors. The first place, with four publications, is occupied by Wang and Lee, with three articles occupying the next position, four authors. These data indicate that only

Tab. 1: Main information.

Description	Results
Time span	1996:2023
Sources (journals, books, etc.)	118
Documents	169
Annual growth rate %	13.83
Document average age	6.27
Average citations per doc	15.09
References	235
Keywords plus (ID)	1,748
Author's keywords (DE)	589
Authors	568
Authors of single-authored docs	19
Single-authored docs	20
Coauthors per doc	3.63
International coauthorships %	2.367
Article	169

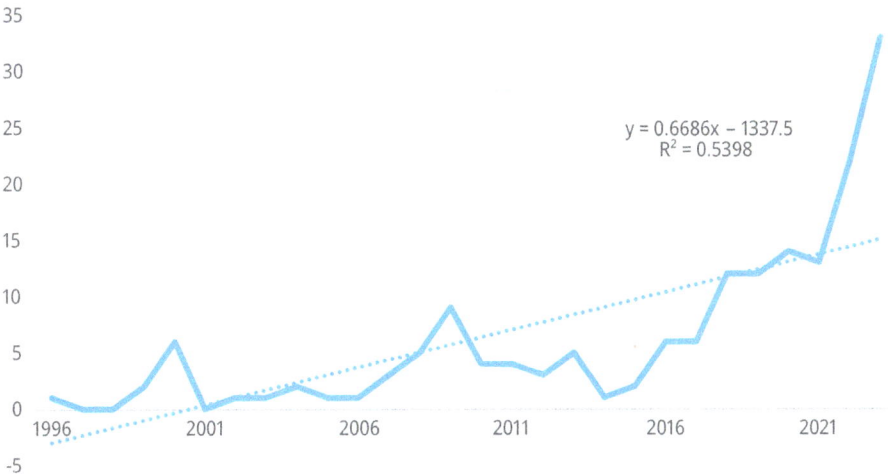

$$y = 0.6686x - 1337.5$$
$$R^2 = 0.5398$$

Fig. 2: Growth in scientific production.
Notes: y: estimated value of scientific production per year. R^2: coefficient of determination or percentage of the total variability of y that can be explained by applying the regression equation.

some authors head the bulk of the research. However, if we consider these authors' total number of citations, Chen is far above the rest, reaching 353, almost seven times more citations than the next author Wang. Chen focuses his research on evaluating the impact of artificial intelligence (AI) in education. Of these authors with the highest number of publications, Lee and D'Andrea stand out with the oldest publications dating back to 2000.

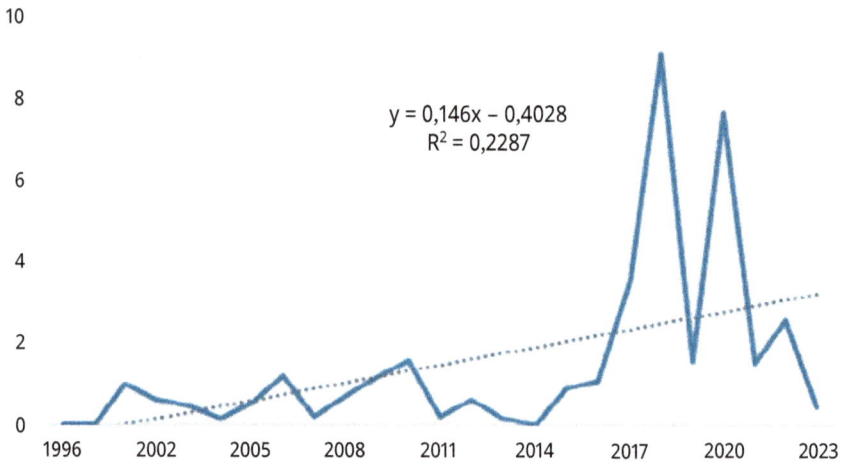

Fig. 3: Average citations per year.
Notes: *y*: estimated value of scientific production per year. R^2: coefficient of determination or percentage of the total variability of *y* that can be explained by applying the regression equation.

Tab. 2: Most cited authors.

Authors	h_index	g_index	m_index	TC	NP	PY_start
Wang	3	4	0.188	56	4	2009
Lee	2	3	0.080	15	4	2000
Chen	3	3	0.600	353	3	2020
Li	2	3	0.667	49	3	2022
Verner	2	3	0.111	14	3	2007
Wu	2	3	0.286	28	3	2018
Ahlgren	2	2	0.111	7	2	2007
Chen	2	2	0.154	21	2	2012
Cheng	2	2	0.118	11	2	2008
D'andrea	2	2	0.08	37	2	2000

The most prolific authors are shown in Fig. 4, which can also be seen in the years they have presented the most outstanding productions. Authors Lee and Wang are the authors with the highest number of publications, with a total of four. The annual output shows how Lee and D'Andrea have been pioneers in publishing articles in this research area since they published their first article in 2000, where the authors discuss the RoboCup project, created to teach systems engineering concepts and practices to students to enable them to design, integrate, and maintain highly complex systems along with the interplay between AI, dynamics, and control theory. It was not until years later that the rest of the authors began publishing. Lee is also the author with the most extended time horizon of publications.

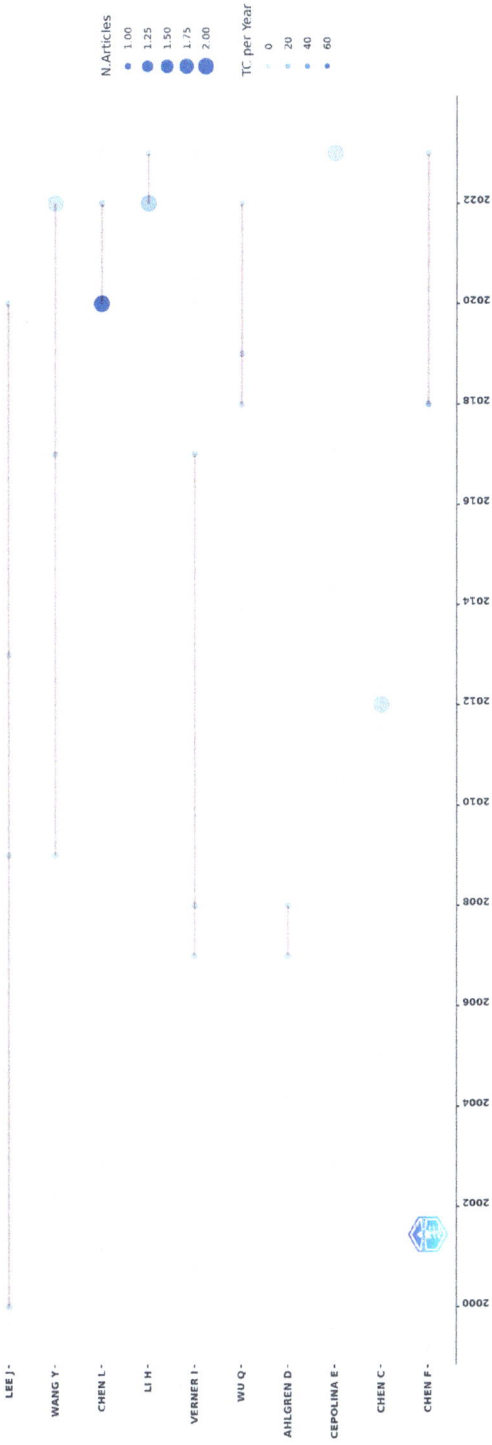

Fig. 4: Authors' production over time.

The most relevant affiliations by number of publications correspond to King Saud University (Riyadh, Saudi Arabia) with six articles and Swarthmore College (Pennsylvania, USA) with five articles. This same institution presents the first publications, with a total of five in 2000. King Saud University shows the highest growth, ranking first in 2023 with six publications and a significant increase in the last 3 years.

The relevance of the journals where the articles analyzed in this study have been published is addressed using the h-index to calculate the scientific productivity of a source based on the number of articles published and their citations. Considering this index, the journal *Robotics and Autonomous Systems* is in the first position, the source with the second highest number of publications related to intelligent robots in education. Looking at the production of the primary sources over time, *Robotics and Autonomous Systems* published six articles in 2008. They led in several publications until 2022 when *IEEE Access* took the lead with eight publications (see Fig. 5). Most journals in which the articles under analysis have been published are in computer science, AI, and robotics.

Table 3 shows the 10 articles with the highest number of citations. Chen et al., with their article "Artificial intelligence in education: a review" from 2020, with 326 citations. In this research, the authors determine that AI has been widely adopted and used in different forms in education, particularly by educational institutions. Starting from AI, it took the form of computers and computer-related technologies, moved on to web-based and online intelligent educational systems, and, ultimately, with the use of embedded computer systems, along with other technologies, the use of humanoid robots and chatbots. In the second place but with fewer citations is the article by Li et al. with 148.

As a complement to enrich the results of the analysis techniques applied in the bibliometric studies, network analysis and clustering have been used [24]. Regarding the cooperation networks generated, the 169 publications involve 32 countries/regions, 305 institutions, and 568 authors. The analysis focuses on the co-country/region network and the coauthor network. Collaborative networks show how authors, institutions, and countries relate to each other in a specific field of research. Figure 6 shows the countries with publications related to intelligent robots and education. The regions represented in darker colors have a higher number of publications.

Specifically, China and North America have dominant scientific productions, with 47 and 30 publications, respectively. Collaboration between countries is very scarce, with a collaborative network linking the USA with China, which has jointly published two papers, and Portugal with Slovakia, which has one joint work.

Figure 7 represents the coauthorship network. The nodes show working groups with fewer self-authors (out of the ten clusters, eight have three or fewer authors). Thus, a close or frequent cooperation network still needs to be possible.

Notably, there needs to be more collaboration between countries and institutions (Fig. 8). Only four countries have conducted joint research. USA has collaborated twice with China and once with India, while Portugal has collaborated with Slovakia.

The keyword analysis is represented through the word cloud, co-occurrence analysis, thematic map, and the evolution of the subject matter. The author's word cloud

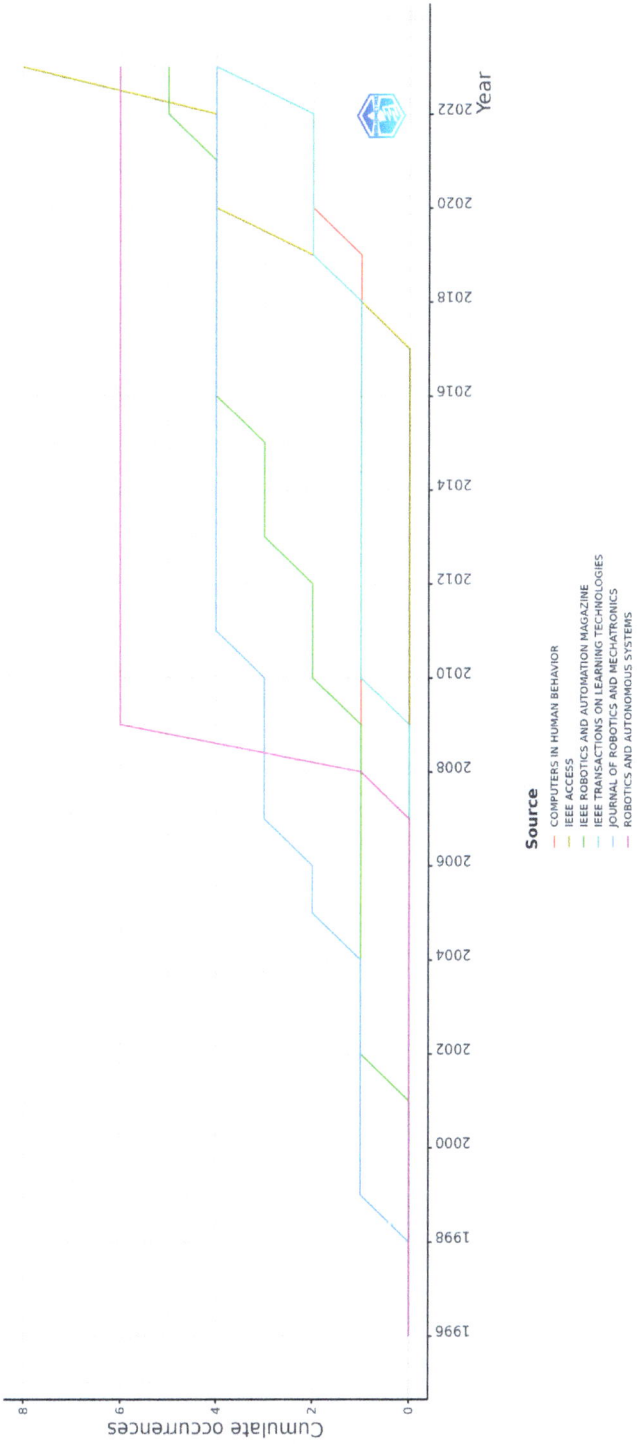

Fig. 5: Source production over time.

Tab. 3: Most cited papers within scientific production/output.

Ranking	Authors	Title	Year	Citations
1	Chen et al.	Artificial intelligence in education: a review	2020	326
2	Li et al.	Reinforcement learning of manipulation and grasping using dynamical movement primitives for a humanoidlike mobile manipulator	2018	148
3	Caruso L.	Digital innovation and the fourth industrial revolution: epochal social changes?	2017	117
4	Du et al.	Online robot teaching with natural human-robot interaction	2018	73
5	Leo et al.	Who gets the blame for service failures? Attribution of responsibility toward robot versus human service providers and service firms	2020	65
6	Edwards et al.	Why not robot teachers: artificial intelligence for addressing teacher shortage	2018	64
7	Liang et al.	Fear of autonomous robots and artificial intelligence: evidence from national representative data with probability sampling	2017	63
8	Maxwell et al.	Integrating robotics research with undergraduate education	2000	58
9	Murphy et al.	Human-robot interaction	2010	51
10	Papadopoulos et al.	A systematic review of the literature regarding socially assistive robots in pretertiary education	2020	51

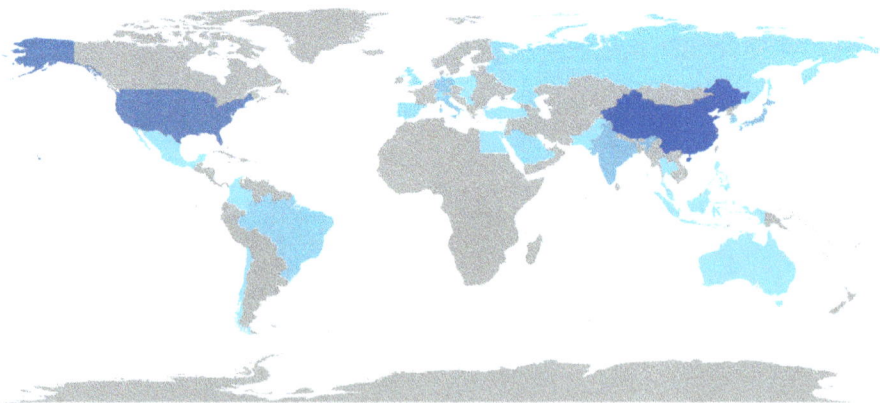

Fig. 6: Countries' scientific production.

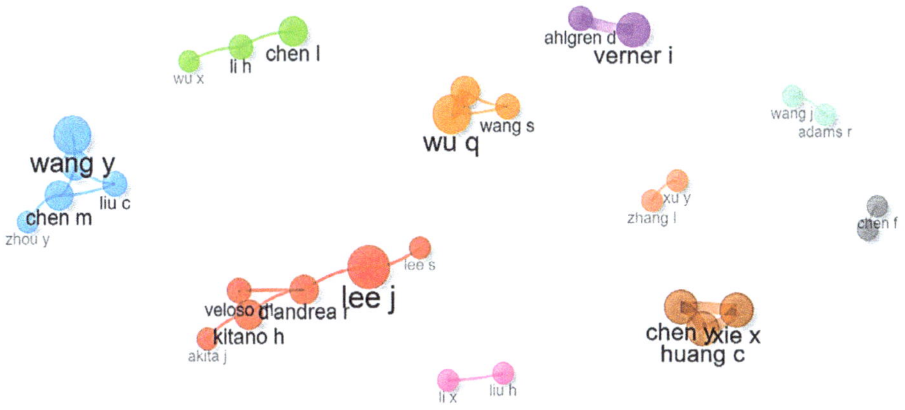

Fig. 7: Collaboration network – authors.

Fig. 8: Collaboration network – countries.

shown in Fig. 9 shows the keywords of the articles selected by the authors. The most frequently occurring word is "intelligent robots," with 146 occurrences. It is followed by "robotics" with 59 occurrences.

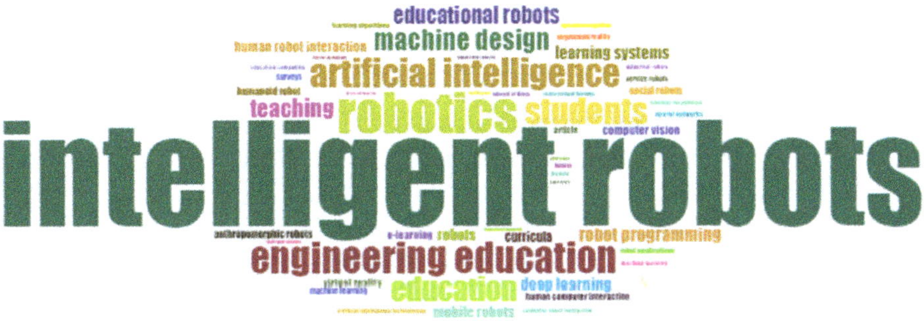

Fig. 9: Word cloud.

In the co-occurrence analysis, three groups of words are distinguished, although only two are representative. These contain words related to technology and education, the most frequent being those related to new technologies and engineering. The third group needs to be more representative, with only three terms, compared to the 23 and 22 words that comprise the two main clusters. The co-occurrence graph in Fig. 10 represents the relationships of lexical units (words) in the analyzed documents. Each word is a node, and the co-occurrence of words in different articles determines the weight of each link. The most prominent words (higher frequency) with larger nodes are located in the central part of the graph, showing the relationship between the two main clusters. The network shows the strength of links between words such as "intelligent robots" and "robotics" or "artificial intelligence" within the red cluster. And strongly related to "engineering education" or "students" in the blue cluster.

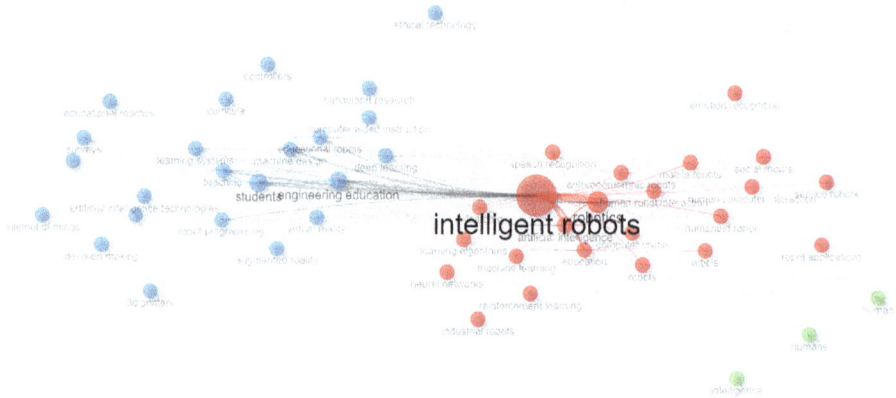

Fig. 10: Co-occurrence network.

The thematic map is shown in Fig. 11. This map is represented in two dimensions and divided into four quadrants, which depend on the density or strength and centrality or interaction of the different terms. In the upper right quadrant with driving themes (high density and centrality) are terms such as augmented reality or the internet of things. The lower right quadrant, with low density and high centrality, represents the core topics "intelligent robots," "robotics," and "engineering education." The upper left quadrant with high density and low centrality shows the niche topics, such as digitalization or architecture, and finally, the lower left quadrant with low density and centrality shows the emerging issues in this case with a reduced number of terms and words like "information and communication technologies."

The thematic evolution is represented graphically through a flow chart with the evolution of the terms from the first articles in the nineties of the last century to the present. In this way, it is possible to analyze the development of the area and conduct research related to intelligent robots in education. It is interesting to note how some

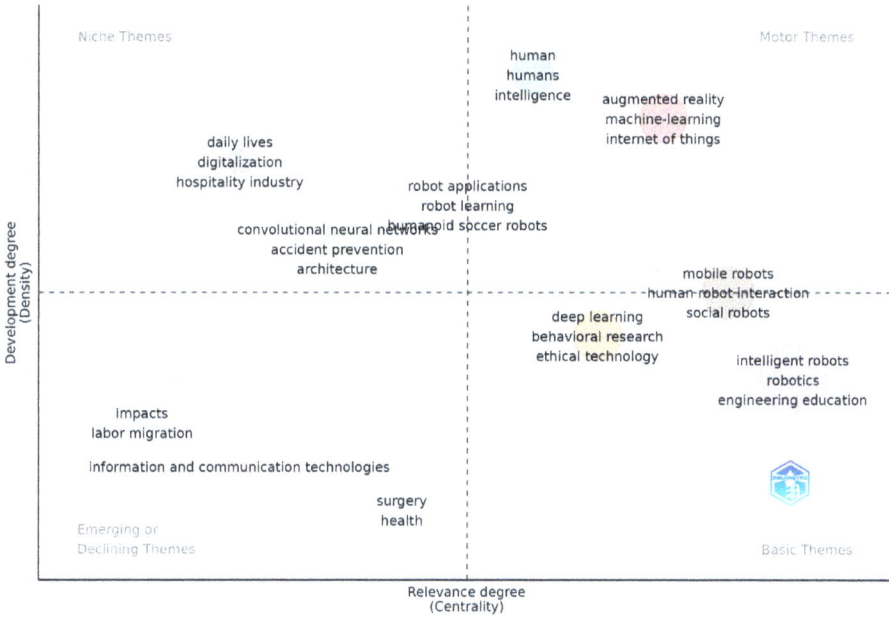

Fig. 11: Thematic map.

terms are merged or divided into several topics. Figure 12 presents the alluvial diagram, divided into three sub-periods of study, each represented in a column horizontally connecting upstream and downstream groups.

The term "intelligent robots" has endured over time, with the transition from the second to the third period seeing the most significant change with the emergence of new related terms. In the period 2017–2021, new terms appear that do not originate from a theme of the previous sub-period, such as "ethical technology" and "architecture." In the rest of the cases, the themes flow from earlier periods. The number of topics has grown in each subperiod, with the highest number in the last period.

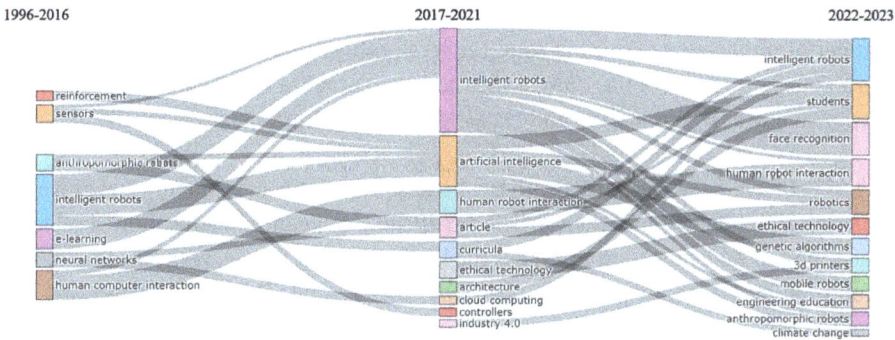

Fig. 12: Thematic evolution.

With the in-depth analysis of the results, we will be able to answer the research questions posed in this work so that it can achieve the main objective proposed in this chapter.

5 Discussion

The results obtained from the bibliometric analysis allow one to determine the current state of the scientific literature related to intelligent robots and education and how this scientific activity has evolved. The bibliometric indicators provided information on the production dynamics, establishing trends and suggesting future research direction. The research questions have been answered with an in-depth analysis of the results, achieving the primary objective proposed in this article.

By in-depth analysis of the results, we have answered the research questions and achieved the primary objective proposed in this article.

RQ1: Which authors are at the forefront of scientific production from the related scientific literature? The 169 articles analyzed were written by 568 authors, with a coauthorship rate of 3.68 authors per work. Single-authored articles accounted for only 11% of the 169 total, with 19 authors publishing individually, only 3%. International coauthorship could be higher, accounting for 2% of the analyzed documents. The authors with the highest number of publications are Wang and Lee with four publications each. Four authors with three publications followed this. Three-quarters of the authors have published only one article, showing the need for more explicit references, according to the number of publications in this area of research. The results clearly show a need for more in-depth and, in many cases, longitudinal research into this educational resource rather than a one-off approach by the researchers who study it. However, the number of citations is considered. In that case, Chen, with 353, is at the top, obtaining a 0.6 in the m-index, an appropriate index in this research when comparing researchers with different durations in their research trajectories. As pioneers in the publication of articles in this area of study, with publications in 2000, are Lee and D'Andrea. Concerning affiliation, the most relevant in terms of the number of publications are King Saud University (Riyadh, Saudi Arabia) and Swarthmore College (Pennsylvania, USA), which together account for 6.5% of scientific production and are the pioneering and fastest-growing institutions, respectively.

RQ2. What have been the works with the highest impact on subsequent related work? It has been highlighted that their article "Artificial intelligence in education: a review" concerning AI and education in 2020 is the highest impact article, with 326 citations. It is followed by 148 citations by the article by Li et al. [35]. This result shows that despite the relative topicality of educational robots, some works have already become benchmarks in the related literature.

RQ3. What are the main topics investigated, which countries lead this scientific production, and which keywords are the most used in the scientific literature analyzed? According to the analysis, topics related to new technologies, such as artificial or augmented intelligence and education, are the main ones. Regarding the thematic evolution, the central term of this research, "intelligent robots," has endured over time, with new related terms such as "ethical technology" and "architecture" appearing. This shows how, despite the different nuances with which the reality of educational robots is connected, the associated terminology has remained unchanged over time, which undoubtedly favors related research. Thirty-two countries have been identified with publications in the research area. The leading countries in production are China and North America, with 47 and 30 publications, or 44% of the total publications. Only four countries have published more than 10 articles, with 90% having fewer than ten publications. The most frequently occurring word is "intelligent robots," with 146 appearances. It is followed by other words within the same topic, such as "robotics" or "artificial intelligence," and others related to education, such as "engineering education" or "students."

RQ4. What conceptual, intellectual, and social structures were generated from the scientific literature analyzed? This has been answered using the corresponding conceptual, intellectual, and social analyses. Three clusters concerning keywords have been identified in the intellectual structure, although only two are relevant. One cluster encompasses terms related to robotics and a second cluster with terms related to technologies in education. The social structure of the collaborative network of authors shows 10 clusters of small size, 80% of which have three or fewer authors. The cluster with the largest number of authors includes six authors. The international coauthorship relationships are reduced, with China and the USA in the lead.

RQ5. What are the future applications of robotic technology in education? Although the current use of intelligent robots in education has become apparent, there is still some way to go. The current use of mathematics to teach concepts, collaboration, and computational thinking should be expanded to other educational fields, encouraging interactive learning [36]. Robots often introduce students to engineering concepts, encouraging interdisciplinary practices and professional development. Although there are positive findings from claims that educational robots improve social skills, more studies are needed to clarify this claim [11]. Considering the growing importance of AI, future research may be related to its role in the educational context [37].

6 Conclusions

The most relevant conclusions can be drawn from the analysis and discussion of the results of this work.

Despite the relative novelty of using this resource in the educational context, a wide range of scientific production addresses the issue of robots in education from different approaches. Notably, the scientific works are of different geographical origins, although it is true that China and the USA stand out over other territories.

An essential part of the scientific works focuses on the fact that robots of an educational nature relate to the reality of ethics and the ethical use of technology. This shows the academy's main concerns regarding using these technological resources in teaching and educational processes.

The connection of new realities such as AI with educational robots is undoubtedly an emerging topic in research and an excellent challenge for the academic community, which the main actors and agents involved must address scientifically and rigorously, given its profusion and growth in such a short period.

The implications of the results of this work for both theory and practice are also important. Regarding theory, this work marks a starting point for future researchers regarding the recurrent concepts connected with using educational robots. Until now, there have been several nuances linked to this reality. Our work provides a turning point in the theory, thanks to the homogenization and unification of some of the most relevant concepts addressed. Likewise, some gaps still existing in the related research were revealed, opening, in many cases, promising lines of related research. On the other hand, the practical contributions of this work are undoubtedly crucial for teachers and professors who use these technologies in teaching their subjects. Considering not only technological aspects, such as emotional and psychological aspects, is also essential for correct use and, above all, for achieving the objectives proposed by the educational institutions.

Regarding future lines of research derived from the results, it should be noted that although this work provides a snapshot of the current state of research on the use of robots in the educational field, it would be advisable in future new works to follow up and study the academic subjects from their birth, evolution, and consolidation.

Another possible future line of research is to focus on the desirability of conducting work using different evaluation methods that would be useful to corroborate the benefits of educational robotics and discover more efficient methods to capitalize on the cultural propensities of students in the context of a robotics curriculum, as well as longitudinal studies that would allow researchers to assess whether there are lasting effects.

References

[1] Zhang Y, Luo R, Zhu Y, Yin Y. Educational robots improve K-12 students' computational thinking and STEM attitudes: Systematic review. Journal of Educational Computing Research, 2021, 59(7), 1450–1481.

[2] Mwangi PN, Muriithi CM, Agufana PB. Exploring the benefits of Educational Robots in STEM Learning: A Systematic Review. International Journal of Advanced Engineering and Technology, 2022, 11(6), 5–11.

[3]   Chiang Y hui V, Cheng YW, Chen NS. Improving language learning activity design through identifying learning difficulties in a platform using educational robots and iotbased tangible objects. Educational Technology & Society, 2023, 26(3), 84–100.

[4]   Rapti S, Sapounidis T. Critical thinking, communication, collaboration, creativity in kindergarten with educational robotics": A scoping review (2012–2023). Computers and Education, 2024, 210, 104968.

[5]   Chu ST, Hwang GJ, Tu YF. Artificial intelligence-based robots in education: A systematic review of selected SSCI publications. Computers and Education: Artificial Intelligence, 2022(July), 3, 100091.

[6]   Jedrinović S, Ferk Savec V, Rugelj J. Innovative and Flexible Approaches to Teaching and Learning with ICT BT. In: Väljataga T, Laanpere M, eds.Digital Turn in Schools – Research, Policy, Practice. Singapore: Springer Singapore; 2019. pp. 171–186.

[7]   Huang R, Tlili A, Xu L, Ying C, Zheng L. Educational futures of intelligent synergies between humans, digital twins, avatars, and robots. Journal of Applied Learning and Teaching, 2023, 6(2), 1–16.

[8]   Benvenuti M, Cangelosi A, Weinberger A, Mazzoni E, Benassi M, Barbaresi M, et al. Artificial intelligence and human behavioral development: A perspective on new skills and competences acquisition for the educational context. Computers in Human Behavior, 2023, 148, July 107903.

[9]   Leoste J, Heidmets M. The Impact of Educational Robots as Learning Tools on Mathematics Learning Outcomes in Basic Education BT. In: Väljataga T, Laanpere M, eds. Digital Turn in Schools – Research, Policy, Practice Singapore: Springer Singapore; 2019. pp. 203–217.

[10] Pellas N. Assessing computational thinking, motivation, and grit of undergraduate students using educational robots. Journal of Educational Computing Research, 2023, 0(0), 1–25.

[11] Atman-Uslu N, Yavuz GÖ, Koçak-Usluel Y. A systematic review study on educational robotics and robots. Interactive Learning Environments, 2023 Dec, 31(9), 5874–5898.

[12] Li S, Liu Z, Qiu M, Huang J, Zheng J, Ding G. Examining the effects of communication features of educational robots on students' cognitive load, attitudes, and learning performance. Journal of Educational Computing Research, 2024, 0(0), 1–26.

[13] Ramírez-Montoya MS, Baena-Rojas JJ, Patiño A. Educational Robotics and Complex Thinking: Instructors Views' on Using Humanoid Robots in Higher Education BT. In: Balogh R, Obdržálek D, Christoforou E, eds. Robotics in Education. Proceedings of the RiE 2023 Conference. Cham: Springer Nature Switzerland; 2023. pp. 117–128.

[14] Pivetti M, Di Battista S, Agatolio F, Simaku B, Moro M, Menegatti E. Educational robotics for children with neurodevelopmental disorders: A systematic review. Heliyon, 2020, 6(10), e05160.

[15] Yang QF, Lian LW, Zhao JH. Developing a gamified artificial intelligence educational robot to promote learning effectiveness and behavior in laboratory safety courses for undergraduate students. International Journal of Educational Technology in Higher Education, 2023, 20(18), 1–31.

[16] Gavrilas L, Kotsis KT, Papanikolaou M-S. Assessing teacher readiness for educational robotics integration in primary and preschool education. Education, 3–13 2024, 3(13), 1–17.

[17] Li S, Yang B. Personalized education resource recommendation method based on deep learning in intelligent educational robot environments. International Journal of Information Technologies and Systems Approach, 2023 Apr, 16(3), 1–15.

[18] Kerimbayev N, Nurym N, Akramova A, Abdykarimova S. Educational robotics: Development of computational thinking in collaborative online learning. Education and Information Technology, 2023, 28(11), 14987–15009.

[19] Tzagkaraki E, Papadakis S, Kalogiannakis M. Exploring the Use of Educational Robotics in Primary School and Its Possible Place in the Curricula BT. In: Malvezzi M, Alimisis D, Moro M, eds. Education in & with Robotics to Foster 21st-Century Skills. Cham: Springer International Publishing; 2021. pp. 216–229.

[20] Martínez-Comesaña M, Rigueira-Díaz X, Larrañaga-Janeiro A, Martínez-Torres J, Ocarranza-Prado I, Kreibel D. Impact of artificial intelligence on assessment methods in primary and secondary education: Systematic literature review. Revista de Psicodidáctica, 2023, 28(2), 93–103.

[21] Saad D, Verner I, Rosenberg-Kima RB. Validation of Teachers' Competencies for Applying Robotics in Science Education BT. In: Balogh R, Obdržálek D, Christoforou E, eds. Robotics in Education. Cham: Springer Nature Switzerland; 2023. pp. 27–35.

[22] Xie H, Zhang Y, Wu Z, Lv T. A bibliometric analysis on land degradation: Current status, development, and future directions. vol. 9, Land. MDPI AG; 2020.

[23] Aria M, Cuccurullo C. Bibliometrix: An R-tool for comprehensive science mapping analysis. Journal of Informetrics, Internet 2017, 11(4), 959–975. Available from https://doi.org/10.1016/j.joi.2017.08.007

[24] Donthu N, Kumar S, Mukherjee D, Pandey N, Lim WM. How to conduct a bibliometric analysis: An overview and guidelines. Journal of Business Research, Internet 2021(April), 133, 285–296.

[25] Nobanee H, Al Hamadi FY, Abdulaziz FA, Abukarsh LS, Alqahtani AF, AlSubaey SK, et al. A bibliometric analysis of sustainability and risk management. Sustainability, Internet 2021, 13(6), Available from https://www.mdpi.com/2071-1050/13/6/3277

[26] Zupic I, Cater T, Čater T. Bibliometric methods in management and organization. Organizational Research Methods, Internet 2015, 18(3), 429–472. Available from https://doi.org/10.1177/1094428114562629

[27] Durieux V, Gevenois PA. Bibliometric indicators: Quality measurements of scientific publication. Radiology, Internet 2010 Apr 8, 255(2), 342–351. Available from https://doi.org/10.1148/radiol.09090626

[28] Xie H, Zhang Y, Duan K. Evolutionary overview of urban expansion based on bibliometric analysis in Web of Science from 1990 to 2019. Habitat International, Internet 2020, 95, 102100. Available from https://www.sciencedirect.com/science/article/pii/S0197397519304527

[29] Farrukh M, Shahzad IA, Meng F, Wu Y, Raza A. Three decades of research in the technology analysis & strategic management: A bibliometrics analysis. Technology Analysis and Strategic Management, Internet 2021, 33(9), 989–1005. Available from https://doi.org/10.1080/09537325.2020.1862413

[30] Kokol P, Blažun Vošner H, Završnik J. Application of bibliometrics in medicine: A historical bibliometrics analysis. Health Information and Libraries, Internet 2021, 38(2), 125–138. Available from https://onlinelibrary.wiley.com/doi/abs/10.1111/hir.12295

[31] Ab Rashid MF, Aziz MAA. A comprehensive overview of world mapping analysis research trends on impact of artificial intelligence in tourism from 2000 to 2022: A Literature Review and Bibliometric Analysis. ICRRD Quality Index Research Journal, 2022, 3(3), 136–160.

[32] Moral-Muñoz JA, Herrera-Viedma E, Santisteban-Espejo A, Cobo MJ. Software tools for conducting bibliometric analysis in science: An upto-date review. Profesional de la Información, 2020, 29, 1–20.

[33] Caputo A, Kargina M. A user-friendly method to merge Scopus and Web of Science data during bibliometric analysis. Journal of Marketing Analytics, Internet 2022, 10(1), 82–88. Available from https://doi.org/10.1057/s41270-021-00142-7

[34] Anwar S, Bascou NA. A systematic review of studies on educational robotics. Journal of Pre-College Engineering Education Research, 2019, 9(2), 19–42.

[35] Li Z, Zhao T, Chen F, Hu Y, Su C-Y, Fukuda T. Reinforcement learning of manipulation and grasping using dynamical movement primitives for a humanoidlike mobile manipulator. IEEE/ASME Transactions on Mechatronics, 2018 Feb, 23(1), 121–131.

[36] Licardo JT, Domjan M, Orehovački T. Intelligent robotics – a systematic review of emerging technologies and trends. Electronics, Internet 2024, 13(3), Available from https://www.mdpi.com/2079-9292/13/3/542

[37] Chen X, Xie H, Zou D, Hwang G-J. Application and theory gaps during the rise of Artificial Intelligence in Education. Computers and Education: Artificial Intelligence, Internet 2020, 1, 100002.

Hızır Dinler

Intelligent Educational Robots in Early Childhood Education: Building the Future

Abstract: This chapter explores the transformative potential of intelligent educational robots (IERs) in early childhood education, focusing on their ability to foster creativity, personalize learning, and promote inclusivity. It provides a comprehensive list and classification of commercially available IERs, emphasizing their diverse functionality and age appropriateness. The authors highlight the importance of integrating IERs to promote a holistic and inclusive learning experience. This includes incorporating science, technology, engineering, and mathematics and maker culture into the curriculum using IER capabilities for personalized learning, and ensuring equal opportunities for children with diverse learning needs, especially those from disadvantaged backgrounds or with learning disabilities. The chapter acknowledges potential challenges, including equitable access to resources and the need for sophisticated software development to meet diverse learning needs. It advocates for a balanced approach that integrates technology while maintaining a connection to reality, and fostering meaningful, impactful learning experiences. Ultimately, the chapter envisions a future where IERs are instrumental in transforming early childhood education and providing children with equitable access to quality education.

Keywords: Intelligent educational robots, artificial intelligence, early childhood education, STEM, maker education, educational technology, robotics in education

1 Introduction

As sociologist and philosopher Zygmunt Bauman states, the world we live in today is "fluid and mechanical." This fluidity is driven by technological advancements and the constant evolution of lifestyles [1]. The world has become a global village where people live in close proximity, yet simultaneously feel distant. We hold the world at our fingertips, yet our interactions, even with those beside us, often resemble separate universes, illuminated by different screens. This duality offers immense opportunities but also threatens essential human experiences like intimacy, sacrifice, and passion. Beyond fluidity, our modern world possesses a mechanical dimension, reminiscent of science fiction imaginings. This dimension permeates every aspect of our lives, from clothing and living spaces to education and social norms.

Corresponding author: Hızır Dinler, Department of Early Childhood Education, Kilis 7 Aralık University, Turkey, e-mail: hizirdinler@kilis.edu.tr

https://doi.org/10.1515/9783111352695-010

Before delving into the role of intelligent robots in early childhood education, it is crucial to understand their nature and origins. Intelligent robots can be viewed as dreams materialized – constructs of metal, circuits, and software imbued with the ability to learn and teach within a world of 0s and 1s.

In his 1714 work, *Monadology*, Leibniz argued against the concept of the mind as a machine, but acknowledged the possibility of machines possessing mental faculties – thinking machines [2]. The question of whether machines can have the ability to think was explicitly addressed for the first time in the historical process by Alan M. Turing in 1950. Alan M. Turing, in his 1950 article "Computing Machinery and Intelligence," begins with the sentence, "I propose to consider the question whether machines can think" [3]. In 1959, Cahit Arf asked the question, "Can a machine think?" again in public conferences [4]. As a result of scientific studies on artificial intelligence (AI), the first concrete steps were taken years later. The most striking of these was Deep Blue, an IBM-developed software program that played chess with Russian chess grandmaster Gary Kasparov in 1997, defeating Kasparov, marking the first human defeat against a computer. In the years that followed, similar encounters created the impression that AI could be a formidable opponent against humans [5]. Similarly, in October 2015, AlphaGo, a program developed by Google's DeepMind to play Go, became the first computer program to beat Fan Hui, a professional Go player, on a 19 × 19 board, without giving him an advantage in the European Go Championship [6]. While these developments initially caused people to have negative feelings, like a dystopian work of fiction, they also began to reveal how a ghost in a machine could communicate with a human being.

On November 30, 2022, ChatGPT, the intelligent AI robot from OpenAI, suddenly entered our social and academic life and reached 1 million users in 5 days [7]. Today, it has more than 100 million active users. ChatGPT, which is trained on very large amounts of data, provides the output requested by the user on any subject based on the input, in a way that is difficult to distinguish from human production, and in the fastest form. This intense interest has grown incredibly fast, and different companies have successively launched AI chatbots with their own language models. Bard (now Gemini), a conversational text generation application from Alphabet, the Google group of companies, is based on LaMDA [8]. Bard/Gemini is based on Meena, a different language model than LaMDA [9]. LaMDA is another AI project by Google [10].

Galactica, the dialog-based text generation application of the Meta group of companies, which owns applications such as Facebook and Instagram, is one of them. Galactica, Meta's dialog-based text generation application, was launched in January 2023. However, it was retired in November 2023 due to ethical concerns and the generation of erroneous information [11]. In addition, Microsoft later integrated ChatGPT 4, called Copilot, first in the Edge browser and then in the Windows operating system as a result of its protocols with OpenAI [12]. Other browsers, such as Brave and Opera, have integrated natural language processing applications such as Llama 2, Claude, and Mistral AI into their AI search applications such as Leo and Aria [13]. The accessibility initiated by

OpenAI quickly diversified, and first the visualization software DALL-E came into our lives, and then applications that could create short videos with SORA started to go public [14]. In this process, independent developers have used existing AI software to create vacation advisors, personal chatbots, voice transcription, illustrating children's books, creating comics, writing stories [15], digitizing personal data, making movies, drawing manga, writing novels [16], and embedding AI characters [17] (NPCs – non-player characters) into games [18]. In addition, large companies such as Amazon are working on AI-enabled customer service, knowledge platforms such as Khan Academy are working on virtual trainers [19], and design applications such as Canva, Midjourney, StableDiffusion, and Adobe are working on AI-enabled editing tools [20].

Amid these advancements, AI is making significant inroads in education. AI-powered educational technologies are poised to revolutionize modern education by providing children with unique and enriching learning experiences. These technologies have been used to enhance child engagement and cater to diverse learning needs. Compared to traditional educational approaches, AI-powered tools offer several benefits, including increased motivation for creative activities, reduced fear of embarrassment or shame, support for personalized learning, and assistance with meaningful teaching tasks.

In recent years, AI has made significant progress in everyday life. In addition to these topics, there are also various developments in the education sector. AI educational technologies will play an important role in modern education by providing children with unique learning experiences and enhancing their learning. These technologies have been used to increase interaction with children and meet a variety of educational needs. AI educational technologies provide a number of benefits to children and teachers compared to traditional education. For example, they increase children's motivation for creative activities, help them avoid embarrassment and shame, and provide teachers with personalized learning and support for meaningful teaching tasks. The educational process involves people acquiring lasting knowledge as well as lasting thinking skills. Research shows that AI for children contributes to the development of important skills, including creativity, emotion control, collaborative inquiry, literacy, and computational thinking, as well as areas such as AI, machine learning, computer science, and robotics [21].

To understand the role of intelligent educational robots in early childhood education, five main elements can be emphasized:
- Fostering creativity: Encouraging the development of creativity in young children.
- Integrating science, technology, engineering, and mathematics (STEM) and maker education (ME) into the curriculum: Incorporating the design and production phases of intelligent educational robots into educational programs.
- Holistic learning experience: Providing a comprehensive learning experience in the digital age through personalized learning support.

- Inclusive and equitable education: Using intelligent educational robots to promote inclusion and equity in education.
- The future of intelligent educational robots in early childhood education: Envisioning the future landscape of IERs in early childhood education.

Let us now focus on these elements.

2 The power of IERs to unlock creativity in early childhood education

Today, the integration of technology into early childhood education is becoming increasingly common. The potential of intelligent educational robots is considered an important tool to unleash children's creativity and enrich their learning experiences [22].

First of all, let's talk about the word "robot," a term first used in 1921 by Czech writer Karel Čapek in his satirical work "Rossum's Universal Robots." In this work, robots are referred to as biologically based servants that do things humans do not want to do [23]. According to Čapek, this word was derived by his brother Josef from the Slovak word for "work" or "labor." Previously, the word 'automaton' was used instead of robot, but after this work, it took its current usage. Again, the three laws that the American chemist Prof. Dr. Isaac Asimov put forward in his story "The Stationary Loop," which forms a part of his book *I am Robot* in 1942, about robots, which he frequently mentions in his works of fiction, are as follows: Robots are cybernetic organisms that cannot harm human beings or watch them being harmed, obey orders, and are programmed to protect their own existence [24]. The predictions of the famous science fiction writer and scientist Asimov are still valid today. Moreover, the law of three robots was later challenged by Isaac Asimov when robots became more complex. In his 1985 novel *Robots and Empire*, he wrote that advanced robots would prevent all of humanity from being harmed rather than a single human being. He called this the "Zero Robot Law." Today, robots continue to exist in our world, as Asimov said, to prevent all humans from being harmed, and to make them better [25].

The power of intelligent educational robots to unleash creativity in early childhood education has emerged as an important research topic. This is a critical time when individuals develop basic skills and explore the world. Research shows that AI tools, including intelligent robots, can increase social interaction among children and lead to greater engagement in learning activities. However, more information is needed on the impact of these AI tools in early childhood classrooms.

PopBot, designed at the Massachusetts Institute of Technology, is a social robot created for young children to learn AI in a constructive, creative, practical, and inexpensive way [26]. Lee et al. [27] developed a robot called Mero and Engkey, programming it to teach letters and words to primary school children. It is observed that there is a positive increase in the speaking skills of these children [27]. Conti et al. [28] examined the effect of learning-supportive robots used for education on kindergarten children. In the study, 81 kindergarten students were allowed to spend time with a humanoid robot telling a story. As a result of the study, it was determined that the robots were attention-grabbing and motivating for young children, facilitated their memorization, and the robot's gestures and mimics that were similar to human behaviors were positively received by the children [28]. In their study, Amanatiadis et al. [29] provided imitation training to children with autism spectrum disorder using humanoid robots named NAO to improve their social behavior. At the end of the study, it was determined that children's social behavior could be improved. Humanoid figures are quite common in intelligent educational robots, and humans are more sympathetic to objects that they can relate to. In this respect, many of the intelligent educational robots we will encounter in the near future will have a face and facial expressions like the humanoid robots that already exist [29].

Chen et al. [30] found that specialized educational tools such as natural language processing technology for language education, as well as AI applications such as educational robots and educational data mining for performance prediction, can have a positive impact on education. Prentzas [31] stated that abstract educational technology provides children with a unique educational experience. However, the integration of educational technology in early childhood education is a more recent trend than in other levels. This fact creates the need to develop, implement, and study application resources and methodologies, specifically for young children. From this point of view, the field of intelligent educational robots needs to be further developed. In this respect, it is a very bright field of study.

Kewalramani et al. [32] examined whether and how technologies such as AI toys in a home-based environment can support children with diverse needs socially and emotionally through play. Based on the concept of "emotional capital" and using a design-based research approach, the study was inspired by parents' intentional use of robotic toys during COVID-19 lockdown periods in 2020 to engage their children with additional diverse needs in home-based play experiences. The study found that the use of AI robotic toys, in combination with physical and artificial environments, positively supports children's social-emotional development and provides an opportunity to build children's emotional capital.

In terms of supporting creativity, intelligent robots are a tool that allows unlimited repetition. In this respect, it is seen that the interest in AI research in the field of education is increasing day by day, and the effects of AI based tools are shared in the form of sample applications in social media posts. However, not enough effort has yet been made to integrate intelligent learning and teaching technologies into educational environments. It is emphasized that these techniques are partial rather than fully integrated and are rarely used on the initiative of the individuals and institutions that are given the opportunity.

Furthermore, since the use of this new technology of AI is not yet standardized, there are reports that special AI tools are being designed for children. However, since these tools are not yet available for use, supervision and guidance issues are very important.

The power of intelligent educational robots to unleash creativity in early childhood education is based on their potential to support and enrich children's learning experiences through play. Thanks to their interactive features, these robots can interact naturally with children and actively engage them in the learning process. Moreover, since intelligent educational robots have programmable features, they can be customized to suit children's learning needs and levels. This offers the possibility to personalize and make children's learning experiences more effective. On the other hand, intelligent educational robots can offer interactive games and activities to help children develop critical thinking and problem-solving skills. In this way, children can both unleash their creativity and have the opportunity to develop important cognitive skills [33].

The power of intelligent educational robots to unleash creativity in early childhood education has the potential to enrich children's learning experiences and contribute to their cognitive development [34]. However, according to the study by Dinler and Cevher-Kalburan [35], it is seen that children's phonological awareness improves with a poetry-focused supportive education program. In this respect, it was revealed that if disadvantaged children are supported, they show more significant language development compared to their peers, and that there may be positive results in terms of equal opportunity and inclusion when focusing on disadvantaged groups. It is seen that this and similar supportive studies can complete the existing deficiencies with the support of technology and intelligent educational robots and provide a better future for children.

However, careful planning, effective implementation, and continuous evaluation are required to fully realize this potential. In the future, the role of intelligent educational robots in early childhood education will become even more important and will continue to make children's learning experiences more effective and enjoyable.

3 Beyond the traditional classroom: innovative approaches for IER-STEM-ME

The traditional teacher-to-child approach to education in the classroom is now outdated. A more interactive and child-oriented education approach is capable of meeting the requirements of the age and within an egalitarian perspective. In this case, an educational method that will positively affect children's lifelong success will be very useful to capture the spirit of the time. This method is called the STEM education approach [36].STEM education aims to integrate these four fields and provide children with knowledge and skills in these areas. STEM education aims to provide children with skills such as problem-solving, critical thinking, analytical thinking, collaboration, and creative thinking. These skills are among the most necessary skills for the next centuries [37].

STEM education, starting in early childhood, enables children to explore scientific and mathematical concepts, learn to use technology, and become familiar with engineering principles. In this way, children who are aware of the principles of digital literacy, media literacy, and even fact-checking will become individuals suitable for twenty-first-century skills. STEM understanding is experienced by children through games, discoveries, experiments, projects, and interactive activities. STEM education in early childhood specifically supports children's sense of curiosity and natural desire to explore. Through play and interaction, children explore simple scientific concepts, develop basic math skills, use technology, and learn engineering principles. This learning is important to develop not only knowledge but also understanding [38].

Children who receive STEM education in early childhood explore scientific processes and problem-solving strategies. For example, during this education, children conduct simple experiments, make observations, form hypotheses, collect data, and analyze and share their results. This process develops their scientific thinking skills and increases their self-confidence. STEM education also develops problem-solving skills in early childhood. By giving them conscious problems, they are enabled to create learning processes with deliberate problems. Children learn to solve problems using technological tools, apply engineering principles, and generate alternative solutions. This process develops children's critical thinking skills and creativity.

STEM education aims to provide children with skills such as problem-solving, critical thinking, analytical thinking, collaboration, and creative thinking. It seeks to develop children's creative thinking skills and encourage innovation. It supports children in generating their own ideas, showing flexibility in problem-solving processes, and finding new solutions. STEM education enables children to use technology effectively and adapt to technological developments. It teaches children computer programming, coding, digital design, and other technology skills. As a result, STEM education aims to provide children with basic knowledge and skills in science, technology, engineering, and mathematics, develop their analytical and critical thinking skills, encourage creative thinking, and teach them to use technology effectively. In this way, STEM education prepares children for the technology-oriented and competitive world of the future [39].

STEM education promotes fundamental knowledge and skills in science, technology, engineering, and mathematics, as well as an in-depth understanding of these fields. It provides children with basic knowledge of science and mathematical concepts, as well as the opportunity to use technology, become familiar with engineering principles, and gain practical experience in these areas [40]. STEM education encourages children to understand scientific processes, use technology effectively, apply engineering principles, and develop mathematical thinking skills. It provides children with analytical and critical thinking skills to solve real-life problems.

One of the main elements of STEM education is ME. ME is an approach to learning that provides children with practical experiences to develop their manual skills, problem-solving abilities, and creativity. This approach focuses on children actively learning by creating products with their own hands, designing, coding, or prototyping. This approach aims to foster their interest in technology and innovation, encourage them to find creative solutions to future problems, and encourage them to pursue careers in STEM fields. Therefore, ME is becoming increasingly important in today's digital age and is being embraced by educational institutions [41].

ME also allows children to direct their own learning and work on projects based on their interests. ME is often based on scientific and engineering principles and is closely related to STEM fields [42]. However, it is not limited to these fields; it can also focus on areas such as art, design, and crafts. Therefore, ME has a broad interdisciplinary approach and aims to provide children with skills in different areas [43].

ME usually involves children learning various skills such as technology use, coding, electronics, 3D design and printing, robotics, woodworking, sewing, and ceramics. Children are given opportunities to design and realize their own projects using these skills. ME also allows children to develop skills such as critical thinking, problem-solving, collaboration, communication, and leadership. Children have to find creative solutions to overcome the challenges they face while working on their projects and, in the process, increase their self-confidence and motivation [44].

ME often involves interactive and practical learning methods such as workshops, project-based learning, design-based learning, and experience-based learning. This approach encourages children to actively participate in the learning process and build their own learning [45]. Innovative strategies and approaches are needed to move beyond the traditional classroom environment, to keep pace with transformational changes in education, and to enable children to learn in a powerful way. In this context, the combination of intelligent educational robots and ME will offer unique opportunities in early childhood education.

Intelligent educational robots and ME are innovative strategies used in early childhood education. This approach offers children the opportunity to develop their manual skills, increase their problem-solving abilities, and encourage creative thinking skills [46]. One innovative strategy is to integrate intelligent educational robots into maker activities. In this way, we can increase children's interactions with technology and make their learning experiences more engaging. Furthermore, the use of intelligent educational robots in ME enhances children's ability to use technology creatively, while at the same time, strengthening their collaboration and communication skills. This strategy increases children's self-confidence and enables them to actively participate in the learning process [47].

Intelligent educational robots and ME are innovative approaches used in early childhood education. This approach provides children with the opportunity to develop their creative thinking and problem-solving skills while teaching them how to use technology effectively [48]. However, for this strategy to be successfully implemented, educators need to develop innovative strategies that understand children's needs and provide them with appropriate learning environments. In the future, intelligent educational robots and ME will be used more in early childhood education, enriching children's learning experiences by increasing their interaction with technology.

4 Enhancing education with IERs: technology-assisted adaptive and personalized learning activities

Early childhood education is a critical period that forms the foundation of important skills that individuals will rely on throughout their lives. In this period, it is vital to provide appropriate and effective learning environments for children's cognitive, emotional, and social development. Traditional educational approaches are limited in their ability to support these goals and adapt to the present day [49]. However, when planned correctly, the use of intelligent educational robots in early childhood education offers new possibilities for enriching learning experiences and unlocking children's potential [50].

The full list of intelligent educational robots that are available for purchase and use today is as follows in alphabetical order:

– **Aibi pocket robot**: The Aibi pocket robot is a pint-sized personal companion that packs a powerful punch. Designed for on-the-go fun and exploration, this pocket-sized bot boasts an impressive array of sensors. Cameras, microphones, and speakers allow Aibi to see, hear, and interact with the world around it. Plus, with a user-friendly smartphone app, you can control Aibi's every move and unlock its full potential.

  For more information, visit https://living.ai/aibi

– **Anki (Digital Dream Labs) Cozmo**: Developed by Anki, Cozmo is an interactive and programmable toy robot. Cozmo is equipped with AI and image recognition technologies. This cute robot can interact with users, show facial expressions, play various games, and be controlled by voice commands. Cozmo can also be programmed with the Cozmo SDK, a Python-based programming environment.

  For more information, visit https://ddlbots.com/products/cozmo-robot

– **Anki (Digital Dream Labs) Vector**: Another robot developed by Anki, Vector is an AI-powered robotic toy similar to Cozmo. Vector is designed as a domestic helper. It can be controlled by voice commands, show facial expressions, sense its surroundings, and perform various tasks. Vector also offers a programming environment that allows users to develop their own skills.

  For more information, visit https://ddlbots.com/products/vector-robot

– **Bee-Bot/BlueBot**: Bee-Bot and BlueBot are simple and fun programmable robotic toys designed for children. These robots are used to help children develop basic coding skills. Bee-Bot and BlueBot receive simple commands via buttons and act accordingly. They are widely used for education, especially at the preschool and primary school levels.

For more information, visit https://tts-international.com/primary/computing-ict/bee-bot-blue-bot-pro-bot-ino-bot

– **Blue-Bot programmable floor robot**: Blue-Bot is a line of programmable robotic toys designed for children. These robots receive simple commands with the help of buttons and act accordingly. Blue-Bot is widely used for education at the preschool and primary school levels to help children develop basic coding skills.

For more information, visit https://tts-international.com/blue-bot-bluetooth-pro grammable-floor-robot/1015269.html

– **Botley robot**: Botley is a programmable robotic toy designed for children, developed by Learning Resources. This robot is designed to help children develop basic programming skills. Botley can be programmed for a variety of activities, and games can be played.

For more information, visit https://learningresources.com/shop/collections/botley

– **Clementoni**: Clementoni is a brand that produces educational robotic games and activities for children. Clementoni's products include robotic education sets and games for various age groups. These sets allow children to develop STEM skills and learn basic coding and engineering concepts.

For more information, visit https://en.clementoni.com/collections/scientific-toys

– **Codey Rocky**: Codey Rocky is a programmable robotics education set developed by Makeblock. This set allows children to learn programming, robotics, and AI. Codey Rocky includes visual programming tools and AI features.

For more information, visit https://education.makeblock.com/product/codey-rocky

– **COJI**: COJI is a programmable toy robot developed by WowWee. This robot is designed to help children develop basic programming skills. COJI can express various emotions using colorful facial expressions and reacts accordingly. It can also be programmed using a simple coding language and can play various games.

For more information, visit https://wowwee.com/coji

– **Colby robotic mouse**: Developed by Learning Resources, Colby is a programmable toy mouse designed for children. This mouse is designed to help children develop basic programming skills. Colby can perform various tasks by moving through a maze and can be programmed using a block-based coding language.

For more information, visit https://learningresources.com/item-code-gor-robot-mouse-activity-set

- **Cubelets**: Cubelets is a programmable robotic toy set developed by Modular Robotics. Cubelets consist of small robotic blocks with different functions that can be magnetically assembled to create a variety of robots. Cubelets allow children to explore basic robotics and engineering concepts.

  For more information, visit https://modrobotics.com/cubelets-for-home

- **Cubetto**: Cubetto, developed by Primo Toys, is a simple and fun programmable robotic toy designed for young children. Cubetto is controlled with the help of buttons and performs various tasks by moving on a plane. This robot helps children develop basic programming and logic skills.

  For more information, visit https://primotoys.com

- **Dash & Dot**: Dash & Dot, developed by Wonder Workshop, is a programmable robotics education set designed for children. Dash is a moving robot, and Dot is a stationary robotic toy. This set allows children to learn basic coding skills and explore STEM subjects.

  For more information, visit https://makewonder.com/en/dash

- **Doc** is a programmable robotic toy developed by Clementoni. This robot is designed to help children learn about programming and robotics. Doc can be programmed using an educational talking coding language and can perform various tasks. It can also show facial expressions and perceive its surroundings through various sensors.

  For more information, visit https://nordics.clementoni.com/products/doc-educa tional-talking-robot-sw-fi

- **Edison robot**: Developed by Microbric, Edison is a programmable and cost-effective robotics education platform. Edison is equipped with various sensors and motors and allows children to develop basic programming and engineering skills.

  For more information, visit https://microbric.com/products or https://meetedi son.com

- **EducatStemBox**: EducatStemBox, developed by Educat, is a programmable robotics education set designed for children. This set includes a variety of activities to help children develop STEM skills. EducatStemBox allows children to learn basic programming and engineering concepts.

  For more information, visit https://educatstembox.com

- **Eilik**: Eilik is a programmable robotic toy designed for children. It can sing, dance, play games, and even connect to intelligent home devices, making it a fun and engaging companion. Eilik doesn't need constant attention. When left alone, it will express itself through spontaneous actions and maintain a standby mode for quick reactivation.

  For more information, visit https://store.energizelab.com/products/eilik

- **Elegoo smart robot car with camera**: Elegoo is a brand that produces programmable robotic education sets designed for children. Elegoo's products include robotic games and activities for various age groups. These sets allow children to develop STEM skills and learn basic coding and engineering concepts.

For more information, visit https://elegoo.com/products/elegoo-smart-robot-car-kit-v-4-0

– **Evo and Bit robots**: Evo and Bit, developed by Ozobot, are small programmable robots. These robots are designed to help children develop basic programming skills. Evo and Bit can follow colored lines and play a variety of games.

For more information, visit https://ozobot.com

– **Finch robot**: Finch is a programmable robotics trainer developed by Carnegie Mellon University. It is designed to help students develop basic programming and engineering skills. Finch can be used for various activities and projects.

For more information, visit http://finchrobot.com / https://ri.cmu.edu/research/robots

– **Gilobaby voice-controlled robotics**: Gilobaby is a company that produces programmable robot toys that offer a fun and engaging way for children to develop STEM skills. Their wide range of products includes various robots that cater to children of all ages and interests. Gilobaby robots are ideal for preschool and elementary school-aged children, suitable for individual or group use, can be used at home or in the classroom, and make STEM education fun and engaging.

For more information, visit https://gilobaby.com/collections/all

– **iRobot Root Rt0**: The iRobot Root Rt0 is a coding and exploration robot created by iRobot. Designed for use in both homes and schools, this robot is an ideal tool for teaching children coding skills. The Root Rt0 uses a visual block-based programming language, making it easy to program even for children with no coding experience. The Root Rt0 has a variety of sensors and LEDs to engage children. It can sing, dance, and play games. The Root Rt0 helps develop STEM skills, as well as problem-solving, creativity, and critical thinking skills. The Root Rt0 comes with a variety of accessories that can be used to expand its capabilities and provide new learning opportunities.

For more information, visit https://irobot.com/en_US/root.html

– **KIBO robotics kit**: Developed by KinderLab Robotics, KIBO is a programmable robotics education kit designed for children. KIBO connects the arts with computer science and coding robotics – without the need for screen time. With KIBO, educators bring lesson plans to life, while kids use their own creativity and self-expression to bring KIBO to life. They investigate their world, trade and share discoveries, develop their computational thinking and STEM skills, and solve problems.

For more information, visit https://kinderlabrobotics.com/

– **KUBO robot**: KUBO is a programmable robotics education set developed by KUBO Robotics. This set is designed to help children develop basic programming and engineering skills. KUBO is a screen-free coding robot designed to introduce computer science and coding to students as young as four years old. KUBO uses a unique system of color-coded tiles called "TagTiles" that snap together to program the robot. These tiles allow children to give KUBO simple commands such as moving forward, turning left or right, and playing sounds. As children progress in their coding skills, they can use more complex combinations of tiles to create more intricate programs.

For more information, visit https://kubo-robot.com/meet-kubo/

– **Lego Mindstorms Education**: Developed by Lego, Mindstorms Education is a robotics education set that helps children and young people develop STEM skills. Lego Mindstorms includes robots made with Lego pieces as well as programming software. With this set, students can build their own robots, program them, and develop various projects.

For more information, visit https://education.lego.com/en-us/downloads/mind storms-ev3/software

– **Makeblock**: Makeblock is a brand that produces programmable robotic education sets designed for children. Makeblock offers a wide range of robotic kits, catering to different ages, interests, and experience levels, starting from beginner-friendly sets like mBot Ranger for outdoor exploration. They often incorporate various electronic modules like sensors and actuators, allowing users to build robots with diverse capabilities like obstacle avoidance, line following, and light and sound interaction. These sets allow children to develop STEM skills and learn basic coding and engineering concepts.

For more information: https://www.makeblock.com/pages/product

– **Marty:** Developed by Robotical, Marty is a programmable humanoid robot specifically designed to enhance STEM learning in educational settings. This educational tool incorporates various features to foster engagement and develop critical skills in students of all ages.

For more information, visit https://robotical.io/about/all-about-marty

– **Matatalab coding set**: Matatalab is a programmable robotics education set designed for children. This set is designed to help children develop basic programming and engineering skills. Matatalab prioritizes a hands-on learning experience. Children use colorful coding blocks to control a friendly robot, MatataBot, eliminating the need for screens and complex text-based coding languages. Children can build various sequences with the coding blocks, instructing MatataBot to move forward, turn, play sounds, and even dance. This gamified approach keeps young minds engaged and motivated.

For more information, visit https://matatalab.com/en/coding-set

– **Max & Tobo coding robot**: Max & Tobo is a programmable robotics education set designed for children. Max & Tobo is an interactive coding robot set designed to take kids ages six and up on an intergalactic coding adventure. This unique system combines the excitement of storytelling with the power of hands-on learning, making coding fun and accessible for young minds. Max & Tobo uses a captivating narrative: Lost robot Max needs your child's help to navigate his way back home through a series of coding challenges. This engaging story keeps children motivated and invested in learning. Max & Tobo utilizes Blockly, a user-friendly block-based coding language. Instead of complex text commands, children drag and drop colorful blocks to control Max's movements, sounds, and actions. This visual approach makes coding easy to grasp, fostering a sense of accomplishment as kids see their programs come to life.

For more information, visit https://www.tutobo.com

– **mBot Ranger**: mBot Ranger is a programmable robotics education set developed by Makeblock. This set includes a variety of activities to help children develop STEM skills. The Makeblock mBot Ranger robot building kit is a 3-in-1 educational robotics kit designed for children ages eight and up. It provides a fun and engaging way for kids to learn about coding, robotics, and engineering while building their own robots. It can be used on all terrains, is perfect for exploring different environments, and has a powerful motor, large wheels, and a gripper arm that can be used to pick up objects. It features a sleek design, low-friction wheels, and a variety of sensors that can be used to avoid obstacles.

For more information, visit https://makeblock.com/pages/mbot-ranger-robot-building-kit

– **Mechabau EVO**: The Mechabau EVO is a versatile STEM educational set designed to spark creativity and ignite passion for building and robotics in children. This set includes a variety of activities to help children develop STEM skills. The EVO set boasts a variety of high-quality parts, including gears, wheels, beams, and connectors. This extensive selection allows children to build a wide range of robots with diverse functionalities. The EVO program offers an optional motorization kit. This addition allows children to bring their creations to life, adding another layer of excitement and engagement to the building experience.

For more information, visit https://mechabau.com/evo

– **MeeperBOT**: MeeperBOT is a programmable robotic toy designed for children. This robot is equipped with various sensors and motors and helps children develop basic programming skills. MeeperBOT utilizes your smartphone or tablet as the control center. Download the user-friendly MeeperBOT app, and transform your smartphone into a powerful coding hub. MeeperBOT champions accessibility with block-based coding. This intuitive system allows children to program their creations using drag-and-drop code blocks, eliminating the need for complex text commands.

For more information, visit https://www.demco.com/meeperbot-2-0-power-platform-for-brick-block-building-with-remote-control-app

– **Miko**: This little companion is powered by AI, making it a fun and interactive friend for children. More than just a toy, Miko develops a life-like personality through inter-action. It can tell stories, play games, and even have conversations, fostering a sense of connection with your child. Miko goes beyond entertainment and offers a gateway to learning. It provides access to educational content, interactive quizzes, and even live online classes, all disguised as fun activities. Miko can be a valuable tool for parents. It can make video calls, allowing you to check in on your child remotely. Parental con-trols also ensure a safe and age-appropriate experience.

For more information, visit https://miko.ai

– **Mind Designer robot**: Mind Designer is a programmable robotics education set de-veloped by Clementoni. This set includes a variety of activities to help children de-velop STEM skills. Mind Designer can be programmed to allow children to learn basic programming and engineering concepts.

For more information, visit https://en.clementoni.com/collections/scientific-toys

– **Mio Robot**: Mio Robot is a programmable toy robot designed for children. This robot is equipped with various sensors and motors and helps children develop basic programming skills. Mio Robot can avoid obstacles and follow your hand. You can use the microphone to command it by simply clapping your hands, while the push-button panel on the robot's back can be used to program its route. The free app has two play sections: coding, to learn the principles of block-based program-ming, and real-time, to control movements, sounds, and light effects. Lastly, with the marker holder and magnet holder, respectively, you can make the robot draw or program it to search for small coins and paper clips.

For more information, visit https://en.clementoni.com/products/mio-the-robot

– **mTiny robot**: mTiny is a programmable robotics education set developed by Make-block. The mTiny robot provides the easiest way to learn coding and mathematical thinking through themed maps, storybooks, game cards, and coding cards. The tap pen, themed maps, and coding cards are designed for learning music, math, lan-guage, and coding. mTiny is a little robot with an adorable panda look, three types of dress-up, more than ten preset emotions, and 300 + sound effects. This set in-cludes a variety of activities to help children develop STEM skills.

For more information, visit https://makeblock.com/pages/mtiny-robot-toy

– **Osmo Coding Awbie**: Developed by Osmo, Coding Awbie is a programmable educa-tional game designed for children. Osmo Coding Awbie utilizes your existing iPad or tablet (base and coding blocks sold separately). This makes it a convenient and ac-cessible learning tool for families who already own these devices. The interactive on-screen environment provides instant feedback. Children can see Awbie follow their code sequence in real-time, allowing them to identify and troubleshoot any errors quickly. This visual reinforcement helps solidify their understanding of coding concepts. This game is designed to help children develop basic programming skills.

For more information, visit https://playosmo.com/en-US/shopping/kits/coding

– **Plobot robot**: Plobot is a programmable robotics education set developed by Plobot through crowdfunding. Command cards are at the core of the Plobot experience, making every interaction tactile and hands-on. Each card represents a block of code. Tap or swipe them on Plobot's head to create a "program" the robot follows. Plobot is a pint-sized coding robot specifically designed to introduce the wonders of coding to young children, ages 4 and up. Plobot's stations act like objective markers for the programming activities in our lesson plan, and turn Plobot into a character in a story. Kids can mold new shapes with clay, attach stickers, or even build crazy formations on its back with Duplo building blocks. They love turning Plobot into their own spaceship! This set is designed to help children develop basic programming and engineering skills.

For more information, visit https://plobot.com

– **Pro-Bot robot**: Pro-Bot is a programmable robotics education set developed by TTS Group. Children can directly program the Pro-Bot's movements using arrow keys on the robot itself. This allows for immediate action and cause-and-effect learning. For a more advanced approach, the Pro-Bot can be programmed using dedicated software on a computer. This software likely utilizes block-based coding, a visual approach that simplifies complex coding concepts into drag-and-drop blocks. This set is designed to help children develop basic programming and engineering skills.

For more information, visit https://tts-group.co.uk/pro-bot-floor-robot-starter-pack/1010501.html

– **Qoopers**: Qoopers is a programmable robotics education set developed by Robobloq. The Qoopers kit provides all the necessary components to build and program six distinct robots: Captain (humanoid), Scorpion (maneuverable robot), Voyager (exploration robot), Guardian (obstacle-avoiding robot), Dozer (powerful robot), and Cavalier (balancing robot). This variety keeps the building and learning experience fresh and exciting. This set includes a variety of activities to help children develop STEM skills.

For more information, visit https://robobloq.com/product/Qoopers

– **Roamer robot**: Roamer is a programmable robotics trainer developed by Valiant. A turtle-inspired educational robot, the Valiant Roamer is designed to introduce children from Pre-K through year 12 to the fundamentals of coding, robotics, and computer science concepts. It is designed to help students develop basic programming and engineering skills.

For more information, visit https://valiant-technology.com

– **Robo Wunderkind**: Robo Wunderkind is a company dedicated to making coding education fun and accessible for children of all ages. They offer a variety of robotic kits and an accompanying app that allows children to build and program their creations, fostering creativity, problem-solving skills, and a love for STEM. Robo Wunderkind uses colorful building blocks that snap together easily. These blocks include lights, motors, sensors, and various structural components, allowing children to build a wide range of robots with diverse functionalities.

For more information, visit https://robowunderkind.com

– **Silverlit**: Silverlit is a toy company that designs and manufactures a variety of robotic toys, including interactive pets, educational coding robots, and remote-controlled robots. Their robots are known for being high-quality, affordable, and engaging for children of all ages. Silverlit offers a variety of robotic pets that can purr, bark, and even walk and follow you around. These interactive pets are a great way to teach children about caring for animals. These remote-controlled robots can perform stunts and tricks. Silverlit also offers a line of coding robots that can be programmed to perform a variety of tasks.

For more information, visit https://silverlit.com

– **Sphero Mini**: Sphero Mini is a small-sized, programmable robotic ball toy developed by Sphero. This compact robot measures about the same size as a ping pong ball. Equipped with a gyroscope, accelerometer, and colorful LED lights, this educational coding robot is more than just a toy. With an interchangeable shell, Sphero Mini helps children develop basic programming skills and can be used for a variety of activities and games. Sphero Mini contains a gyroscope, motor encoders, and accelerometer, as well as a bright, colorful LED to add an extra layer of excitement to the learning process.

For more information, visit https://sphero.com/products/sphero-mini

– **Thames & Kosmos**: Thames & Kosmos is a company that creates educational kits and robotic toys designed to introduce children to the exciting world of science, technology, engineering, and mathematics (STEM). Their robotic kits allow kids to learn by doing, building various robots and then programming them to perform different tasks. This kit is designed for children ages 6 and up. It includes everything you need to build a simple rover that can be programmed using a remote control. The kit also includes a variety of coding challenges to help kids learn the basics of programming.

For more information, visit https://store.thamesandkosmos.com/collections/science-kits?pf_t_subject=Robotics

– **Think & Learn Code-a-Pillar Toy**: Developed by Fisher-Price, the Code-a-Pillar Toy is an educational toy designed for children at an early age. "Code-a-Pillar" programmable caterpillar is a programmable toy for children from 3 to 6 years old. It is made up of eight sections connected by USB sticks that can be easily inserted into each other in order to create an itinerary with the help of symbols that indicate top, left, right, straight ahead, or pause.

For more information, visit https://service.mattel.com/us/productDetail.aspx? prodno=DKT39&siteid=27

– **Thymio robot:** Thymio is a programmable educational robot designed specifically for elementary and middle school students. Equipped with various sensors and LED lights, Thymio helps students learn basic programming skills. Used to develop robotic projects, Thymio offers students the opportunity to explore programming and engineering topics.

For more information, visit https://thymio.org/

– **WeDo:** Developed by LEGO Education, WeDo is a robotics education set designed for primary school students. WeDo includes robots built with LEGO bricks as well as simple programming software. With this set, students explore science and engineering topics while developing basic programming skills. WeDo offers students an interactive and fun learning experience.

For more information, visit https://education.lego.com

– **Yixin Emo Pet:** Yixin Emo Pet is a robot pet designed to develop emotional intelligence and interaction skills. Designed especially for children and teenagers, this robot forms an emotional bond with users and can interact with them in different ways. Yixin Emo Pet helps users strengthen their social and emotional skills.

For more information, visit https://living.ai/emo

– **Zmrobo:** Zmrobo is a programmable robot designed to support STEM education. Using Zmrobo, students learn basic programming skills and develop various robotic projects. This robot offers students the opportunity to explore science, technology, engineering, and math.

For more information, visit http://stemtown.com

The classification of educational robots is a complex task, as diverse criteria can be employed, each offering unique insights. This categorization is not merely an academic exercise; it serves a practical purpose for educators and researchers. By understanding these classifications, users can make informed decisions about which robot best aligns with their specific pedagogical goals, student demographics, and budget constraints. We will delve into five key classification methods.

Classification by brand/manufacturer: This approach groups robots based on the company that created them. This can be a useful way to identify specific models or compare offerings from different manufacturers. For example, grouping can be made according to brands such as Anki (Cozmo, Vector), Bee-Bot/BlueBot, and Clementoni.

Classification by functional characteristics: This method categorizes robots based on their functional characteristics, such as whether they are programmable, configurable, or designed for specific educational purposes. This approach can help users choose a robot that aligns with their learning goals, for example, programmable robots (Dash & Dot, Edison Robot, mBot Ranger), configurable robots (Cubelets, Makeblock), or robots for learning (EducatStemBox, Matatalab Coding Set).

Classification by age group: This classification considers the age of the intended user when grouping robots. This is an important factor to consider, as different age groups have different needs and learning styles, for example, robots designed for children (KIBO robotics kit, The Mind Designer Robot), robots designed for teenagers (Evo and Bit robots, mBot Ranger), or robots for students of all ages (Lego Mindstorms Education, Sphero Mini).

Classification by programming abilities: This approach focuses on the level of programming complexity that each robot supports. Some robots are designed for simple programming and are suitable for young children, while others offer more advanced programming capabilities that can cater to experienced users, for example, robots that can do simple programming for children (the Blue-Bot Programmable Floor Robot, the Botley Robot), or robots that support more advanced programming languages (Lego Mindstorms Education, Sphero Mini).

Classification by physical design: This method groups robots based on their physical form factor. This can be an important consideration for users who want a robot that is esthetically appealing or has specific practical features, for example, robots in animal form (Yixin Emo Pet, The Finch Robot), robots in vehicle form (Max & Tobo Coding Robot, Robobloq), or robots in humanoid form (Eilik, Robo Wunderkind).

Intelligent educational robots support children's learning process in an impressive way. First of all, these robots offer an interactive and fun learning experience. They attract children's attention with their colorful and attractive designs, mobility, and sound effects. With these features, they motivate children's learning process and increase their curiosity.

Furthermore, intelligent educational robots offer personalized learning activities that can adapt to the individual learning needs of children [51]. These robots, which can sense their environment through sensors and can be programmed, can offer customized activities according to children's interests and learning pace. This allows each child to learn at his/her own pace and adapt to the learning process [52].

Intelligent educational robots also develop collaboration and social interaction skills. Through group projects and interactive games, children strengthen their collaboration, problem-solving, and communication skills. In this way, children practice their ability to interact and cooperate socially, while also learning social values such as empathy and sharing. Robots, as the most important helpers of personalized education, will play an important role in the near future in developing a child-friendly education approach, especially in disadvantaged groups.

5 The right to inclusive and equitable education

The Oxford Dictionary defines "inclusion" as the deliberate inclusion of all people, ideas, and perspectives in society [53]. "Inclusive education" is defined as education that is inclusive of all children, especially those with special needs or disabilities [54]. Inclusion is an element that needs to be understood and adapted to daily life for multicultural structures and educational approaches. According to research by Shlapko et al. [55], inclusive education is defined as an educational focus that aims to provide equal opportunities for all children and equal rights for children in need of support by promoting human diversity, as defined by international standards. According to Prerna [56], inclusive education is defined as embracing all children, promoting respect for human dignity, and contributing to the development of a rights-based society.

The right to inclusive and equitable education rests on a fundamental principle: every child, regardless of background or ability, deserves equal access to quality education. This entails not only ensuring equal access to classrooms and resources but also employing inclusive teaching methods and tools that cater to diverse learning needs. In this context, intelligent education robots play an important role in the education system of the future. When it comes to inclusion, the quantity or physical elements of people are not the main focus but the quality of education and accessibility to education, taking into account their positive contribution to the mental process. In this respect, intelligent education robots increase the possibility of providing a similar quality education to everyone in the education process. In cases where there is not even a teacher, intelligent education robots are a comprehensive set of applications that can even carry out an education process without a teacher, providing equal opportunity in education. In AI-supported applications, taking into account all developmental differences, including special needs, and ignoring socio-economic differences, providing equal and high-quality educational support to everyone has the potential to ensure that children exposed to challenging situations or with special needs have similar high-quality learning experiences.

Consider, for instance, the potential impact on children from marginalized communities: students with disabilities who require personalized learning support; children from migrant families navigating linguistic and cultural transitions; girls in communities where educational opportunities are traditionally limited. Intelligent robots, coupled with internet access, can help bridge these gaps and create more equitable learning environments. One of the most important issues at this stage is to train intelligent education robots to create a free curriculum using a constructivist or project-based approach, and with the right prompts, a unique education system can be established in foreign language teaching in any existing conditions, in terms of opening the door to all the necessary knowledge of the world and experiences such as future popularity, job opportunities, and world travel. Intelligent educational robots offer an inclusive educational environment for children with different learning needs. Through customizable learning activities and programming options, each child receives an education that suits their learning pace and style. For example, children with special needs, such as autism spectrum disorder, can have an interactive and calm learning experience through intelligent educational robots.

Another issue of inclusion in the education process is the equal rights of cultural diversity [56]. Today, due to factors such as wars and natural disasters, masses of thousands of people are displaced for a better life, sometimes controlled and sometimes uncontrolled. Even though there are various transition and adaptation education programs and support practices for these refugees, migrants, forced migrants, and people under temporary protection status, new wars, new borders, and unsafe situations for children are frequently encountered every day. The problem of access to education is a potential future problem for every country. Looking at the OECD's (Organization for Economic Co-operation and Development) PISA (Program for International Student Assessment) reports, even if some countries seem to have an education system, there is a daily decline in subjects such as reading comprehension and basic math skills, and in some countries, values far below the average. At this stage, the promotion of cultural diversity and socio-economic policies, rising inflation, and difficult living conditions are forcing people to migrate. Yet, with each new wave of migration, education integration programs are not equally effective. Even changing country policies lag behind these new waves of migration. In this situation, intelligent education robots will be one of the biggest supporters for children in the shadow of war to access an equitable and inclusive education.

Another important issue in the education process is specific learning disabilities. Specific learning disabilities have different subdimensions such as dyscalculia (difficulty in learning mathematics) and dyslexia (difficulty in recognizing characters such as letters and sounds). While in the past, these learning difficulties were considered to be unsolvable, or even a result of mental retardation, today we understand that it is quite natural for bright minds with high potential to have challenging educational processes in some cases. In specific learning disabilities, if the individual is not supported, failure in the educational process will bring about permanent and lifelong problems such as educational phobia, school phobia, social phobia, and social-emotional retardation. However, if the individual is supported according to his/her own learning process, the specific learning disability becomes a minor inconvenience that will only take a little time, and the person will turn his/her potential into kinetic, provided that he/she is taken care of a little, and will have a successful educational life and, more importantly, become a happy individual. In this respect, supporting intelligent education robots according to the personal characteristics of the individual will be a great facilitator.

In addition, intelligent educational robots can also be considered a potential tool for gender equality and increasing women's participation in STEM fields. By combating gender-based stereotypes and prejudices in society, intelligent educational robots can increase girls' interest in STEM subjects and instill in them the confidence that they can succeed in these fields. However, there can also be challenges in using intelligent educational robots to ensure an inclusive and equitable educational environment. For example, differences in access to financial resources can make it difficult to distribute robots equitably across schools and communities. This can deepen inequalities and hinder an inclusive education system.

It is quite clear that intelligent educational robots can be an important tool in achieving the goals of equity and inclusion. So, how can a more inclusive education system exist with future developments? At this stage, it is important for an inclusive education system to have easy and cheap access to universal education tools – robots. There is also a need for innovative approaches and studies on electricity storage – batteries so that the charge of the robots can be used for a longer period of time. Another issue is that we need software that can develop qualified programs, identify existing special conditions (autism spectrum disorder, dyslexia, stuttering, etc.) or problem elements (forced migration, socio-economic disadvantage, etc.), and develop the existing curriculum accordingly. Since AI technology is a system that learns by assimilating everything that exists, it should continue to maintain a high-quality learning process with a strong educational flow against each new situation, taking into account the new differences that will emerge in the near future.

In today's studies, it is seen that intelligent education robots support children's language development and foreign language acquisition. When we look at the studies, the existing intelligent education robots are working on topics such as ME, coding, language learning, math skills, etc. In this respect, it is very important to make the intelligent education robots of this culture more inclusive, to encourage each child with support in the form of an assistant, an educational support for each child, and a digital device that can create its own educational support for each child, and to ensure that everyone can access this support. Admittedly, socioeconomic factors often intersect with and exacerbate existing inequalities in education. To mitigate this, it is crucial to ensure equitable access to intelligent educational robots and provide comprehensive training programs for educators. These programs should equip teachers with the knowledge and skills to effectively integrate these tools in diverse learning environments, fostering inclusion and promoting educational equity.

6 The future of IERs: a new era in early childhood education

Intelligent robots, AI-supported education models, AI-supported education tools, and many more, which we can consider very new today, will help us in our challenging journey for humanity and its future as a huge ecosystem interacting with each other to make the world a better place in the near future [57].

Let's imagine that a child in a disadvantaged area and a child in a private school in a city of skyscrapers are educated with the same curriculum, a thought that may be considered utopian and impossible at the moment. But let's imagine a unique and customizable education system so that every child is given equal opportunity, can repeat as much as they wish, and learn about the world, space, science, technology, engineering, mathematics, biology, physics, literature, art, and similar disciplines with the same up-to-date examples. The sense of equality that this education system will give is very promising for the future of education.

When we look at the OECD reports on education around the world, in some countries, children are taught in purposeful, homework-free classrooms, and even in classrooms without walls, while some schools naturally have no walls [58]. However, our main point is the strengths of the education provided and the processing of the existing ore into something better, that is, into a jewel. Because children who go to school in the forest and children who go to forest schools have the same opportunities, one leads to a disadvantaged future by sharing the existing opportunities from the perspective of poverty and deprivation, while the other ensures the upbringing of self-confident and happy children with plenty of oxygen and free experiences [59]. The inequality here has existed since ancient times and will continue to exist. Even though nongovernmental organizations and nonprofit foundations are fighting meaningfully against this inequality, the inclusive support of technology will be one of the biggest supporters of making the world a more equal place. In particular, child-centeredness, enriched experiences, and interactive interactions will provide the main turning points of the future [60].

After a family picks up their child from a school with intelligent educational robots, they will go to a restaurant and order from a robot chef who recognizes the child and asks for a small exercise from the lessons they have just taken. Then, they will take a driverless taxi to a hobby garden close to the city, water their plants, and return home in the evening. With augmented reality glasses, they will be able to participate in a virtual experience about the Inca Civilization, practice languages with friends from around the world, have a conversation with Shakespeare about tips on the girl they're in love with, and go out to dinner with their family. This may sound like science fiction for now, but in the very near future, large ecosystems will be able to see you and ask if you want your favorite food, without you telling them what it is, and this will become quite normal. Large ecosystems can be nurturing when they are not single-handed. A digital education revolution, inclusive and culturally diverse, with a massive mobilization of learning, can transform the myths of the past into plausible understandings and put a more accessible future in the palms of people's hands.

While intelligent robots are washing the dishes, people can more comfortably play sports, read books, and spend time with each other. It should not be forgotten that the nature of the process determines everything. For instance, the internet, while providing easy access to information in the post-truth era, has equally facilitated the sharing of disinformation and fake information. Now, with developing AI, it will be possible in the very near future to make people speak words they did not say, with deepfake videos; to have people who do not exist, become social media influencers; to see games, inventions, and songs by nonexistent actors; even to go to an exhibition of paintings made by people who do not exist; and participate in many other fantastic and perhaps dangerous experiences. Technology always carries with it a certain amount of danger, but it is the way it is used that determines the process. Deepfake can be used to create political disinformation with a malicious voiceover, or deepfake can be used to have Napoleon Bonaparte visit a classroom and teach children about his battles, his regrets, his childhood, and even his educational process through an AI-powered educational program. It would be a fascinating experience. Children can be taken on a trip to the James Webb telescope, a field trip to the Tiangong space station, or the International Space Station, again with augmented or virtual reality experiences. This would be an unforgettable teaching process for children. How difficult is it to teach history, science, math, literature, or physics to a child who sits, drinks chocolate milk, and talks with Alexander the Great, Suleiman the Magnificent, Plato, Homer, Nikos Kazantzakis, Charles Darwin, Gabriel Garcia Marquez, İhsan Oktay Anar, Marie Curie, or Albert Einstein?

In all this process, one of our main goals should be to keep in touch with reality and to realize ourselves in the age of post-truth and to accumulate meaningful experiences. Reality is a key that opens the human mind. In this regard, while toothaches were once thought to be caused by evil spirits, goblins, and fairies haunting the teeth, we now know that human biology consists of muscles, nerves, and cells that are the output of an evolutionary process. Therefore, while integrating the ancient wisdom of the past into the educational process, a critical perspective is essential.

Whereas in the past, it took hundreds of years for people to digest the cultural changes that made life easier, now, with a new AI tool every day, they are suddenly confronted with outputs at the limits of the mind. In order to catch up with the requirements of the age, our minds must always be open and adaptable. The rest is just enjoying life and producing. It should not be forgotten that man will exist in this world as long as he/she produces.

7 Conclusion

Intelligent educational robots will become one of the most important areas of study in early childhood education in the near future. This book chapter explores the role of intelligent educational robots in early childhood education and examines the benefits and important elements of these technologies. The review reveals that AI educational technologies enrich children's learning experiences and add value to the educational process [61].

First, the importance of intelligent educational robots in fostering creativity was emphasized. While developing children's creative thinking skills is considered a critical element for their future success, intelligent robots can provide effective support in this process. Secondly, the need to integrate STEM and maker culture into the curriculum is emphasized. This approach is geared toward strengthening children's scientific thinking, problem-solving, and technology utilization skills. Third, the importance of a holistic learning experience is emphasized. Meeting children's specific needs with personalized learning support can contribute to the creation of a learning environment that meets the needs of the digital age. Finally, the role of intelligent educational robots in supporting an inclusive and egalitarian education approach is discussed. These technologies are expected to provide equal opportunities for children with different learning styles and needs and contribute to the creation of an inclusive learning environment.

In future studies, in order to gain a deeper understanding of intelligent educational robots and, by extension, AI, it would be beneficial to begin by addressing the underlying issues that contribute to the current state of affairs. These include, but are not limited to, class inequality, the problem of access to basic rights, being in disadvantaged regions, and the reflections of economic crises on education. Scientists have a significant role to play in this regard. It is of great importance that scientists internalize and support existing technologies, making the world a better place. This includes concepts such as equality of opportunity, accessibility, and sustainability. In this regard, regions deprived of education should be provided with independent or state-supported internet access, such as Starlink, or large language model-supported tools with the right hardware support. These regions should have standardized access. Moreover, it is of great importance to support societies facing specific social and economic challenges, to the same extent as efforts to establish colonies on Mars or the Moon, with intelligent educational robots and AI technologies.

Once the issue of the lack of access to basic education has been addressed, it is necessary to direct efforts toward more proactive initiatives. The support process must be diversified and shaped according to the needs of the population in order to prevent the return of previous problems and to address any potential issues that may arise. At this stage, it is essential to identify people's ideas and needs through screening inventories on a regular basis. Once again, scientists have a significant role to play in this process. A proactive, intelligent educational robot support process should be monitored through national or international projects and multicultural support studies. The implementation of studies on different nations and cultures will facilitate the inclusion of all countries in PISA reports and their positioning at the pinnacle of global rankings. This, in turn, will contribute to the improvement of child welfare on a global scale.

It is of vital importance that scientists collaborate in order to develop a high-quality education process and curriculum, as well as to provide effective problem-solving and preventive services. There is a wealth of literature on this subject, yet there has not been a comprehensive study that provides a complete overview. Identifying this issue and disseminating the findings is an essential initial step. Subsequently, the world can be defined through the use of infographics and easy-to-read graphics, which can then be used to facilitate the creation of a universal language of comprehensibility. This can be achieved with the support of large language models. Despite the ease of access to information and resources, many countries continue to demonstrate low levels of success in subjects such as reading comprehension, respect for differences, and basic mathematical perspective. At this juncture, intelligent educational robots should surmount these adverse circumstances through individualized instruction, scientific inquiry, technological innovation, engineering expertise, and mathematical studies. They should also develop a more child-friendly, family-friendly, and teacher-friendly interface that is, in essence, human-friendly.

Today, we understand that in the not-too-distant future, intelligent educational robots will become an indispensable element in early childhood education, just as they are in every aspect of our lives. Currently, various schools of early childhood education are preferred around the world, such as the Waldorf education system [62], the Mindfulness awareness approach [63], and the Montessori education approach and tools [64]. Digital education support, which is placed next to these understandings and systems, first with augmented reality [65] and then with virtual reality, will become available to everyone and become an integral part of our daily education flow, especially with the understanding of inclusivity and equality for disadvantaged groups.

In conclusion, this study on the role of intelligent educational robots in early childhood education highlights the potential of technology in education and provides important clues on how future educational environments may be shaped. It is hoped that advances in this field will help children receive a better education and maximize their potential.

References

[1]   Bauman Z. 44 letters from the liquid modern world. Polity Press; 2010.
[2]   Leibniz GW. Monadology (written 1714). Discourse on Metaphysics and the Monadology, Prometheus Books; 1992.
[3]   Turing AM. Computing machinery and intelligence. Mind, 1950, 59(236), 433–460. https://doi.org/10.1093/mind/LIX.236.433
[4]   Arf C. Can the Machine Think and How Can It Think?. In: Proceedings of the Atatürk University studies in the neighborhood and public education publications conferences series. Erzurum, Türkiye: Atatürk University Press; 1959. 1, pp. 91–103.
[5]   Hsu F. IBM's deep blue chess grandmaster chips. IEEE Micro, 1999, 19(2), 70–81. https://doi.org/10.1109/40.755469
[6]   Gibney E. What Google's winning Go algorithm will do next. Nature, 2016, 531(7594), 284–285. https://doi.org/10.1038/531284a
[7]   Doshi RH, Bajaj SS, Krumholz HM. ChatGPT: Temptations of progress. The American Journal of Bioethics, 2023, 23(4), 6–8. https://doi.org/10.1080/15265161.2023.2180110
[8]   De Vynck G. Google's Gemini's the latest AI software entering fierce competition. The Washington Post, 2023.

[9]   Ray PP. ChatGPT: A comprehensive review on background, applications, key challenges, bias, ethics, limitations and future scope. Internet of Things and Cyber-Physical Systems, 2023, https://doi.org/10.1016/j.iotcps.2023.04.003.

[10]   Lo CK. What is the impact of ChatGPT on education? A rapid review of the literature. Education Sciences, 2023, 13(4), 410. https://doi.org/10.3390/educsci13040410

[11]   Metz C. Human touch on AI has unpredictable outcomes. The New York Times, 26 Sep, 2023, B1.

[12]   Teubner T, Flath CM, Weinhardt C, Van der Aalst W, Hinz O. Welcome to the era of chatgpt et al. the prospects of large language models. Business & Information Systems Engineering), 2023, 65(2), 95–101. https://doi.org/10.1007/s12599-023-00795-x

[13]   Jiang AQ, Sablayrolles A, Roux A, Mensch A, Savary B, Bamford C, et al. Mixtral of experts. arXiv Preprint arXiv:2401.04088, 2024, pp. 1–13.

[14]   Cavalcante DC. Sora: Dawn of digital dreams. Takk™ Innovate Studio, 2024, pp. 1–4.

[15]   Baron NS. Who wrote this?: how AI and the lure of efficiency threaten human writing. Stanford University Press; 2023. https://doi.org/10.1515/9781503637900

[16]   Dobrin SI. AI and writing. Broadview Press; 2023.

[17]   Huang J, Huang K. ChatGPT in Gaming Industry. In: Beyond AI: ChatGPT, web3, and the business landscape of tomorrow. Cham: Springer Nature Switzerland; 2023. pp. 243–269. https://doi.org/10.1007/978-3-031-45282-6

[18]   Zhou W, Peng X, Riedl M. Dialogue shaping: Empowering agents through npc interaction. arXiv preprint arXiv:2307.15833, 2023, pp. 1–6.

[19]   Dicerbo K. Why Not Go All-in with Artificial Intelligence?. In: Stephanidis C, Antona M, eds. HCI international 2021 – late breaking papers: Virtual, augmented and mixed reality; design and user experience; learning and training technologies; assistive technologies; human factors and ergonomics; universal access; cross-cultural design; and user studies. Cham: Springer International Publishing; 2021. pp. 361–369. https://doi.org/10.1007/978-3-030-77857-6_25

[20]   Javan R, Mostaghni N. From canvas to screen: Resurrecting artists of the past. Radiology, 2023, 308(1), e231118. https://doi.org/10.1148/radiol.231118.

[21]   Su J, Yang W. Artificial intelligence in early childhood education: A scoping review. Computer Education Artifical Intelligence, 2022, 3, 100049. https://doi.org/10.1016/j.caeai.2022.100049

[22]   Alrobia NR, Alsaleh N. Educational Robots and Creative Thinking Skills. Humanities and Management Sciences – Scientific Journal of King Faisal University, 2022, https://doi.org/10.37575/h/edu/210080.

[23]   Capek K, Cervera J. RUR (robots universales rossum). Escuelas Profesionales Sagrado Corazón; 1971.

[24]   Asimov I. I robot. Narkaling Productions; 1940.

[25]   Asimov I. Robots and empire. Ballantine Books; 1985.

[26]   Williams R, Park HW, Oh L, Breazeal C. Popbots: Designing an artificial intelligence curriculum for early childhood education. In: Proceedings of the AAAI conference on artificial intelligence. 2019. pp. 9729–9736. https://doi.org/10.1609/aaai.v33i01.33019729

[27]   Lee S, Noh H, Lee J, Lee K, Lee GG. Cognitive Effects of Robot-assisted Language Learning on Oral Skills. In: Second language studies – acquisition, learning, education and technology. Tokyo, Japan; Sept. 22–24, 2010.

[28]   Conti D, Cirasa C, Di Nuovo S, Di Nuovo A. "Robot, tell me a tale!" A social robot as tool for teachers in kindergarten. Interaction Studies, 2020, 21(2), 220–242. https://doi.org/10.1075/is.18024.con

[29]   Amanatiadis A, Kaburlasos VG, Dardani C, Chatzichristofis SA. Interactive social robots in special education. In: 2017 IEEE 7th international conference on consumer electronics-Berlin (ICCE-Berlin). IEEE; 2017. pp. 126–129. https://doi.org/10.1109/ICCE-Berlin.2017.8210609

[30]   Chen X, Cheng G, Zou D, Zhong B, Xie H. Artificial intelligent robots for precision education. Educational Technology & Society, 2023, 26(1), 171–186. https://doi.org/10.30191/ETS.202301_26(1).0013

[31] Prentzas J. Artificial Intelligence Methods in Early Childhood Education. In: Artificial intelligence, evolutionary computing and metaheuristics: In the footsteps of Alan Turing. Berlin, Heidelberg: Springer Berlin Heidelberg; 2013. pp. 169–199. https://doi.org/10.1007/978-3-642-29694-9_8

[32] Kewalramani S, Kidman G, Palaiologou I. Using Artificial Intelligence (AI)-interfaced robotic toys in early childhood settings: A case for children's inquiry literacy. European Early Childhood Education Research Journal, 2021, 29(5), 652–668. https://doi.org/10.1080/1350293X.2021.1968458

[33] Zviel-Girshin R, Luria A, Shaham C. Robotics as a tool to enhance technological thinking in early childhood. Journal of Technology and Science Education, 2020, 29, 294–302. https://doi.org/10.1007/s10956-020-09815-x

[34] Belpaeme T, Kennedy J, Ramachandran A, Scassellati B, Tanaka F. Social robots for education: A review. Science Robotics, 2018, 3. https://doi.org/10.1126/scirobotics.aat5954

[35] Dinler H, Cevher-Kalburan N. The effect of poetry focused supportive educational program to preschool children's phonological awareness. Kastamonu Education Journal, 2021, 29(5), 1034–1051. https://doi.org/10.24106/kefdergi.715624

[36] Kelley TR, Knowles J. A conceptual framework for integrated STEM education. International Journal of STEM Education, 2016, 3, 1–11. https://doi.org/10.1186/s40594-016-0046-z

[37] Erol A, İvrendi A. STEM education in early childhood. Journal of Early Childhood Studies, 2021, 5(1), 255–284. https://doi.org/10.24130/eccd-jecs.1967202151265

[38] Dinler H, Simsar A, Yalçın V. Examining the 21st century skills of 3–6 year old children in terms of some variables. e-International Journal of Educational Research, 2021, 8(2), 281–303. https://doi.org/10.30900/kafkasegt.941467

[39] McDonald CV. STEM education: A review of the contribution of the disciplines of science, technology, engineering and mathematics. Science Education International, 2016, 27(4), 530–569.

[40] Hassan MN, Abdullah AH, Ismail N, Suhud SNA, Hamzah MH. Mathematics curriculum framework for early childhood education based on science, technology, engineering and mathematics (STEM). International Electronic Journal of Mathematics Education, 2019, 14(1), 15–31. https://doi.org/10.12973/iejme/3960

[41] He X, Li T, Turel O, Kuang Y, Zhao H, He Q. The impact of STEM education on mathematical development in children aged 5–6 years. International Journal of Educational Research, 2021, 109, 101795. https://doi.org/10.1016/j.ijer.2021.101795

[42] Hachey AC, An SA, Golding DE. Nurturing kindergarteners' early STEM academic identity through makerspace pedagogy. Journal of Early Childhood , 2021, 1–11. https://doi.org/10.1007/s10643-021-01154-9

[43] McDonald S, Howell J. Watching, creating and achieving: Creative technologies as a conduit for learning in the early years. British Journal of Educational Technology, 2012, 43, 641–651. https://doi.org/10.1111/j.1467-8535.2011.01231.x

[44] Mcclure ER, Guernsey L, Clements DH, Bales SN, Nichols J, Kendall-Taylor N, Levine MH. STEM starts early: Grounding science, technology, engineering, and math education in early childhood. Joan Ganz Cooney Center at Sesame Workshop. Joan Ganz Cooney Center at Sesame Workshop, 2017.

[45] Aronin S, Floyd KK. Using an iPad in inclusive preschool classrooms to introduce STEM concepts. Teaching Exceptional Children, 2013, 45(4), 34–39. https://doi.org/10.1177/004005991304500404

[46] Ebner M, Schön S, Narr K, Grandl M, Khoo E. Learning design for children and youth in makerspaces: Methodical-didactical variations of maker education activities concerning learner's interest, learning with others and task description. Advances in Social Science, Education and Humanities Research, 2021, https://doi.org/10.2991/assehr.k.211212.038.

[47] Daniela L, Lytras MD. Educational Robotics for Inclusive Education. Technology, Knowledge and Learning, 2018, 24, 219–225. https://doi.org/10.1007/s10758-018-9397-5

[48] Lieto M, Inguaggiato E, Castro E, Cecchi F, Cioni G, Dell'Omo M, et al. Educational robotics intervention on executive functions in preschool children: A pilot study. Journal to Computers in Human Behavior, 2017, 71, 16–23. https://doi.org/10.1016/j.chb.2017.01.018

[49] McDonald S, Howell J. Watching, creating and achieving: Creative technologies as a conduit for learning in the early years. British Journal of Educational Technology, 2012, 43, 641–651. https://doi.org/10.1111/j.1467-8535.2011.01231.x

[50] Camilli G, Vargas S, Ryan S, Barnett W. Meta-analysis of the effects of early education interventions on cognitive and social development. Teachers College Record, 2010, 112, 579–620. https://doi.org/10.1177/016146811011200303

[51] Chin KY, Hong ZW, Chen YL. Impact of using an educational robot-based learning system on students' motivation in elementary education. IEEE Transactions on Learning Technologies, 2014, 7, 333–345. https://doi.org/10.1109/TLT.2014.2346756

[52] Ko W, Han JS, Ji S, Nam K, Lee S, Shon W. Development of task planner for u-intelligent educational robots. 2009 ICCAS-SICE, 2009, 5699–5702.

[53] Oxford Dictionaries. inclusion (n.). Accessed March 30, 2024. https://www.oxfordlearnersdiction aries.com/definition/english/inclusion

[54] Oxford Dictionaries. inclusive education (n.). Accessed March 30 2024. https://www.oxfordlearners dictionaries.com/definition/english/inclusive-education

[55] Shlapko T, Filonenko S, Kovalevskyi MM, Hryb A. International experience of legal regulation of inclusive education. Anal Compar Jurisprud, 2022, https://doi.org/10.24144/2788-6018.2021.03.14.

[56] Baria P. Inclusive education: A step towards development of right based society. Journal of Learning and Educational Policy, 2023, https://doi.org/10.55529/jlep.32.37.43.

[57] Cox AM. Exploring the impact of artificial intelligence and robots on higher education through literature-based design fictions. International Journal of Educational Technology in Higher Education, 2021, 18. https://doi.org/10.1186/s41239-020-00237-8

[58] Robertson SL. Provincializing the OECD-PISA global competences project. Globalisation, Societies And Education, 2021, 19(2), 167–182. https://doi.org/10.1080/14767724.2021.1887725

[59] Sobel D. Learning to walk between the raindrops: The value of nature preschools and forest kindergartens. Children, Youth and Environments, 2014, 24(2), 228–238. https://doi.org/10.1353/cye.2014.0035

[60] Li D, Chen X. Study on the application and challenges of educational robots in future education. In: 2020 International Conference on Artificial Intelligence and Education (ICAIE). 2020. pp. 198–201. https://doi.org/10.1109/ICAIE50891.2020.00053

[61] Tanaka F, Kimura T. Care-receiving robot as a tool of teachers in child education. Interaction Studies, 2010, 11, 263–268. https://doi.org/10.1075/is.11.2.14tan

[62] Nicol J, Taplin J. Understanding the Steiner Waldorf approach: Early years education in practice. Routledge; 2012. https://doi.org/10.4324/9780203181461

[63] Zelazo PD, Lyons KE. The potential benefits of mindfulness training in early childhood: A developmental social cognitive neuroscience perspective. Child Development Perspectives, 2012, 6(2), 154–160. https://doi.org/10.1111/j.1750-8606.2012.00241.x

[64] Rothmeyer J. Interaction of environmental education and Montessori pedagogy. 2019.

[65] Suylu M. The effect of the training program for augmented reality applications on the attitudes and opinions of pre-service preschool teachers. [Master's thesis]. Denizli: Pamukkale University Institute of Educational Sciences; 2019.

Barbora Stenová, Karolína Miková, and Lucia Budinská

Interactivity of smart educational robots

Abstract: In this contribution, we analyze the interactivity of smart educational robots in the context of their use in the educational process, as well as their significance for the development of twenty-first-century skills and STEAM. Furthermore, in this chapter, we present three main categories of interactivity: input, output, and input-output. Input interactivity involves sensors that enable robots to gather information about their surroundings. Output interactivity is manifested in the robot's responses to stimuli, such as illuminating lights or playing sounds. Input-output interactivity includes predefined functions where the robot automatically responds to input information. The chapter focuses on the significance and methods of implementing the interactivity of smart educational robots in the educational process and emphasizes their benefits for skill development and student motivation.

Keywords: Smart educational robots, sensors, add-ons, interactivity, programming, STEAM, twenty-first-century skills

1 Introduction

In the current era of digitalization, modern technology has become an essential part of our daily lives. The development of artificial intelligence (AI) has given new capabilities to modern technology, leading to its widespread use in various industries [1]. This has resulted in the creation of autonomous cars, autonomous robots, image/text analytics, and automated customer services, among many others. Advances in modern technology, coupled with AI, will have a profound impact on the requirements for future employment, as well as the necessary competencies for life in this century [2]. Therefore, it is important for students to gain an overview of how the digital world and AI work, as well as the opportunities, challenges, and risks involved. There are a number of ways in which students can be introduced to the principles of AI, for example, through the use of different applications that use AI or software to train it [3]. However, to gain a better understanding of AI, it is recommended to interact with it,

Corresponding author: Karolína Miková, Faculty of Mathematics, Physics, and Informatics, Comenius University Bratislava, Bratislava, Slovakia, e-mail: karolina.mikova@fmph.uniba.sk
Barbora Stenová, Faculty of Mathematics, Physics, and Informatics, Comenius University Bratislava, Slovakia, e-mail: barbora.stenova@fmph.uniba.sk
Lucia Budinská, Faculty of Mathematics, Physics, and Informatics, Comenius University Bratislava, Slovakia, e-mail: lucia.budinska@fmph.uniba.sk

https://doi.org/10.1515/9783111352695-011

which can be achieved through the use of sensors, sound and light cues, or other tools. Educational robotics provides a practical and engaging way to introduce this concept of interaction to students.

In this chapter, we present research focusing on the analysis of the meaning and use of interactivity in intelligent educational robots. Another goal was to analyze the spectrum of interactivity offered by the most commonly available educational robots in the teaching process (e.g., sensors, speakers, lights, or other elements that provide interaction between robots and the external environment).

This chapter is divided into seven parts. The first part of the chapter reviews related works and various studies that focus on integrating the interactivity of intelligent educational robots into the teaching of programming as well as other school subjects. In the second part, we discuss the importance of educational robotics in several countries. We describe the differences in the way educational robotics is taught in Slovakia, Czechia, and Western and Central European countries. We focus on the integration of intelligent educational robots into appropriate school subjects. In the conclusion of this section, we present the contribution and advantages of intelligent educational robots in teaching programming. In the third section, we focus on the great importance of twenty-first-century skills, which are considered key competencies in professional and personal life. We explain how educational robotics contributes to the development of digital literacy, creative learning, and problem-solving in STEAM education. Next, this chapter explores the broader uses of educational robotics, including language learning and support for autistic children, and examines its role in closing the gender gap and improving student performance. The fourth part of the chapter discusses the importance of programming in education, particularly in primary schools, and its role in developing students' logical and algorithmic thinking. We further describe that intelligent educational robots are a more effective tool for teaching programming than child programming languages because they offer immediate and tangible feedback. In the sixth and most extensive section, we present the results of our research. We describe in detail the importance of the interactivity of intelligent educational robots, as well as the range of possibilities of working with sensors and other robot add-ons, which ensure the interaction of robots with the external environment, in the teaching process of programming. In this chapter, we present and also justify the classification of the scope of interactivity that we have developed into three categories: input, output, and input-output interactivity. We define each domain individually and also give examples of sensors and their use in teaching programming. In the last section, we present different perspectives on the significance of our results and the limitations that affected them. Finally, we give an overview of potential projects that could build on the results of this research and aim at a deeper understanding of the different aspects of interactivity and its understanding by learners.

2 Related works

Various studies and projects are currently underway to integrate sensors into the teaching of robotics and other subjects. This trend is supported by a review study [4] in which the authors analyzed several high-level empirical studies on K-12 robotics teaching and learning. Finally, the authors selected a total of 22 studies that they systematically analyzed from different perspectives, one of which was the use of sensors in robotics education. The findings showed that 13 of the 22 studies, representing 59.09%, integrated sensors into their robotics teaching process.

Many school curricula [5–7] for teaching robotics emphasize robot interactivity (working with sensors, sound, and light of intelligent educational robots) as a key concept that students should master. For example, Berry et al. [6] present how pupils can become practically familiar with the sensors, lighting effects, and sounds of intelligent educational robots as part of robotics education. Their curriculum includes activities such as localization and mapping, which students implement through different types of sensors such as infrared sensors (light, proximity, distance), bump/contact/limit switches, encoders, gyroscopes, and global positioning systems. Furthermore, according to this curriculum, students learn to work with sound and lighting effects such as blinking and sound generation based on certain conditions. The sensors in intelligent educational robots function as the primary measurements and also allow the robots to interact with their environment. By applying high-quality and validated curricula to education, students learn to understand, for example, technological systems because they have the opportunity to explore how robots collect and evaluate data using sensors [5].

A review study [4] shows that several educational approaches focus only on superficial robot programming (robot construction, script writing, and robot control). Thus, students do not gain a deeper understanding of the robot hardware and also do not learn in detail about the large range of motor and sensor control options. This is one of the reasons that influenced the authors of the EUROPA project [7], which aims to provide a platform for students to learn about robots from introductory to advanced concepts.

In recent years, several projects [8–11] have emerged that aim to integrate robotic sensors not only into the teaching of programming but also to connect them with other subjects such as physics, biology, chemistry, environmental science, and even art. In 2019, elementary school teachers in America, along with a research team, developed a curriculum [8] to promote the integration of professional learning and computational thinking using sensors. The curriculum includes various activities, such as exploring mold growth conditions or understanding and designing maglev trains using micro:bit sensors. In 2013, the Global Robotics Art Festival (GRAF) [9] was held in Michigan, bringing together the fields of robotics and art. The goal of this festival was to engage students in STEM learning and the arts. One of the requirements for the final robot model was that it had to be programmed by students and integrate sensors. For example, students used a proximity sensor to identify the position of a finger so that the guitar robot knew what note to play, or proximity sensors were used in a robot that created an LED light show based on sensor data. Other projects mentioned focused on environmental issues and the integration of technology, engineering, and computational thinking to improve air quality and water resources. High school students developed an air quality monitoring kit using an Arduino robotic system and a complementary set of programmable sensors [10]. The "Teaching Environmental Sustainability – Modeling My Water District" project allows students to use a variety of sensors to collect data, which are then transmitted to a mobile app via Bluetooth connectivity [11]. These researches and projects clearly show that sensors are not only a tool within robotics education but also a powerful instrument to connect technology education with other disciplines, thus supporting students' holistic learning and development in the modern technological world.

3 The importance of educational robotics

Teaching systems are different in different countries. In some countries, such as Slovakia or the Czech Republic, educational robotics is most often taught in informatics classes, which have taken a prominent place in the educational process. In Western and Central European countries, educational robotics is taught more in block or STEM classes. As we move eastward in Europe, the trend of separating the curriculum into individual subjects becomes more prevalent. Thus, not only teachers but also researchers are dealing with the problem of integrating smart educational robots into appropriate subjects. The most common are informatics and computer education classes, where the curriculum aims to develop computational thinking [12] by programming in different (children's) programming environments. In contrast to these environments, smart educational robots allow students to develop not only computational thinking but also motor skills, spatial imagination, creativity, planning, critical and algorithmic thinking. They also enhance students' soft skills and competencies such as teamwork, creativity, thinking in context, and problem-solving in constructing models and programs [13]. Learning programming and algorithmic thinking becomes more attractive and comprehensible for students thanks to the interactive manipulation with smart educational robots. Educational robots make it easy to build on cross-curricular content, to use innovative methods or playful forms of learning, but the most important factor is the demonstration of a connection between theoretical knowledge and real life, thus improving the motivation and engagement of students in the learning process [14].

4 Twenty-first-century skills and STEAM

As a result of the rapid development of modern technologies and the digital world, the skills that students should acquire during their studies are changing in order to adapt to the rapid changes in their future lives. For this reason, efforts are being made in several countries to change education systems and incorporate modern teaching methods [15]. One of these methods is educational robotics, which also plays an important role in the development of twenty-first-century skills that are considered crucial in today's modern era. Working with educational smart robots contributes not only to the development of digital literacy but also to a lasting learning experience through learning by making. In the same way, activities with robotics lead students to responsibility and independence in problem-solving tasks. Several educational smart robots focus on the development of STEAM knowledge, which combines the teaching of science, technology, engineering, art, and mathematics. These developed competencies are now considered essential for many professions [16]. In addition, educational robotics is finding applications in other fields, such as language education, where smart educational robots assist students in correct pronunciation [17]. It is also used in the education of autistic children, where interactivity plays a key role in the learning process [18]. Smart educational robots are becoming one of the effective tools that prepare students for the challenges of technological advances and the complex demands in their future personal or professional lives. In addition to technical skills, working with educational robotics develops computational thinking and enables students to approach tasks without gender differences, especially seen in the lower grades (at least according to our experience in school practice). In the upper grades, some girls are already fixated on the belief that they will not be as good at this as boys. Some of our newest findings even suggest that underachieving students perform surprisingly well when working with educational robotics, compared to their normal scores. Therefore, it is important that educational robotics has a place in the curriculum as early as the lower grades of elementary school or even in preprimary education [19] and naturally allows students to experience success in a technical area such as engineering and programming.

5 The advantage of robotics over conventional programming

In recent years, programming has become an integral and important part of informatics education, even in primary schools [20]. Programming provides students with a tool to develop logical, algorithmic thinking and twenty-first-century skills [21]. In the process of programming, students are given space to express their thoughts, ideas, and creativity, thus developing their personalities [22]. They also learn to solve complex problems systematically and efficiently. Programming helps students to understand the workings of modern technologies that they face in their daily lives. There are a number of children's educational software programs that use an attractive graphical interface and fun interactive elements to make learning interesting and engaging for students [23]. However, nowadays, when students are exposed to digital technologies in their daily lives, educational software may be less attractive to them. On the other hand, smart educational robots are an exciting tool for students to learn programming and can compete with technologies from children's lives. In 2016, the BBC in the UK distributed one micro:bit device to every student aged 11–12 to support the learning of programming. A year later, Gibson conducted research [24] to explore primary school students' attitudes toward learning programming using the micro:bit. The results showed that students considered working with the micro:bit to be easy, attractive, and beneficial for learning the basics of programming and problem-solving. Students also showed an increased interest in programming. Smart educational robots are interactive physical devices that offer instant feedback in the real world [25], which increases students' interest and motivation. Unlike conventional children's educational software, in which the results of programming are abstract outputs on a screen, robots offer a visible and tangible response for each step of the program. This allows learners to immediately see if their program is performing the tasks correctly, making it more effective in identifying potential errors and deficiencies in their programs [26]. Teaching with smart educational robots is more comprehensible and concrete, thus expanding learning opportunities and stimulating the interest of the younger generation in programming.

6 Interactivity: meaning

Interactivity is the specific element of robotics that separates working with smart educational robots from traditional programming in children's educational software [27]. Interactivity is provided by integrated or added sensors, speakers, and even robot lights that respond to specific stimuli from the environment. This interactive way of communication can be compared to simplified human behavior and the way humans communicate with each other. For this reason, the interactivity of the robots is intuitive for the students and also helpful for their programming and learning. Thanks to the robot's immediate feedback on their programming solution to a problem, students can more easily evaluate the correctness of their solution or detect errors [26]. Teachers and students nowadays see an increased level of autonomous decision-making by robots due to the effective combination of AI and appropriate sensors (autopilot, chatbots, home vacuum cleaner, etc.).

In recent years, our research team has been looking specifically at the interactivity possibilities of smart educational robots. We have analyzed in detail the spectrum of possibilities offered by the robotics kits that are most commonly used in the teaching process, specifically their sensors, such as speakers, lights, or other elements that ensure the interaction of robots with the external environment. The focus was not on the technical features of the sensors but rather on their functionality and effective use in the teaching process. Additionally, the text highlights how students can benefit from using the sensors in programming. Based on a careful analysis of the interactivity of the individual robots, we concluded that this area is so broad that it requires more detailed specification, so we created a categorization consisting of three groups. We created three categories: input, output, and input-output interactivity, which we describe in the following sections.

6.1 Input interactivity

When we talk about input interactivity, we are mainly referring to sensors that gather various information from the environment, based on which the robot can act interactively. There are a number of sensors that a robot can have. However, for our study, we selected only those that were present in most of the educational robots we analyzed. We list them and briefly describe the way in which students can use them. Students can use the **proximity sensor** when evaluating their distance to another object or to detect a change in distance. They can also use **input devices** such as buttons, which are often used because they are easy to operate. There is less speculation about the principle of data collection. These buttons often have preprogrammed functions, or students can program them with an action to be performed by the robot when the button is pressed, such as moving forward, stopping the robot, or turning on a light. Using a **color sensor**, students can recognize the color palette of other objects, which can be used, for example, in an activity with vehicles and traffic lights. Depending on the type of sensor, this can be just basic colors or a wider range of colors. In some cases, the sensor even has a function to determine light or dark colors, which can be used by students to recognize day and night or to program a black line tracing activity. Thanks to the **tilt sensor** (gyroscope), students can experiment with different tilt and rotation changes of the robot. This includes shaking the robot or making it free-fall. This sensor records information about the angle and direction of the robot's tilt, often in the form of a number, which the sensor sends and which students can use further in their programs. Another commonly used input device is a **microphone**, which students use to record their own sounds and to measure the intensity of loud sounds. Students find working with sounds very engaging. Another sensor is a **thermometer** that measures the surrounding temperature. It is not often included in educational robots, rather as an external sensor with the possibility of being connected to the robot. However, students can use this sensor to simulate interesting projects like smart home activities – smart heating and air conditioning. The force sensor detects the force of a squeeze, indicating whether the sensor is compressed or not. Lastly, the **light sensor** captures the light intensity. Students can use it in black line tracking activities or for intelligently switching on lights when it gets dark. Connecting it to the experience of everyday life is very much possible.

6.2 Output interactivity

Another category we created is the output interactivity of smart educational robots, which is demonstrated in the robot's responses to certain stimuli. These reactions can be noticed by the learners using their perceptions. Among the output reactions, the most common are the **lighting** up or **flashing** of a light, or the **playing** of **a sound**. These are the robot's reactions according to commands preprogrammed by the students. Of course, such reactions also include the robot's **movement**, but we have placed this in a separate category, which we will not go into in detail now because it has many different aspects and settings. All these kinds of output interactivity offer an attractive and experiential form of learning that also promotes students' interest in technology. In order to understand what can be meant by general terms, we give specific examples of how students can work with light or sound with smart educational robots.

Smart educational robots usually have multiple built-in lights that students can control independently by entering commands in the software to create interesting lighting effects. Some also have a predefined light effect function, such as a specific or random sequence of flashing lights in different colors. Students can use these effects, for example, when programming a police car, a disco, etc. A special type of light is a light matrix made of several LED lights, which opens up possibilities for students to draw different shapes, symbols, letters, and numbers. Students can simulate simplified displays of devices such as a vending machine, washing machine, or microwave. There are also extension lights for different robots that can be attached to the robot using the appropriate ports. One such example is the colored LED light strip for the micro:bit, which provides different variations of lighting or flashing. Students have the opportunity to experiment with these lights and set them to different light colors or brightness intensities. They enter these settings into the programs either by selecting from a specific color palette or by adjusting the parameters of individual components from the RGB palette, thus achieving a varied color range.

The sound responses of smart educational robots are also very attractive to students because of the wide range of variety. One of the most common sound variations is the playing of a separate tone, for which they can set the pitch, length, or the instrument it is played on. Students can thus program a sound sequence, for example, to create a simple melody.

When working with some robots, students can also use a database of different sound files stored directly in the software they are using. Students use these sounds to support their programmed stories. Often, an editor is also available in the software environment for recording and editing their own sound files.

Some smart educational robots have an interesting sound function that can simulate the sounds of human emotions, such as joy, sadness, boredom, surprise, squeaks, or animal sounds. Of course, the choice of sounds is adapted to the age of the student for whom the educational robot is intended. These sounds can be launched by the students using preprogrammed commands, or the robot performs them autonomously in response to a certain stimulus.

6.3 Input-output interactivity

We have defined this category of interactivity as predefined functions that an intelligent educational robot automatically performs after receiving some input information from sensors and responds with a specific predefined behavior without this behavior being specified by the user (student). This behavior is most often built-in from the factory. We have identified the three most common input/output functions: **line tracking, broadcasting, and lighting** a certain color. These predefined functions simplify the interaction between the robot and the user, providing consistent and predictable robot behavior in different situations. These predefined responses may vary depending on the complexity of the robot.

In the line-following function, the smart educational robot uses a sensor to follow a line on the ground and move along it. In some cases, it also has defined responses at intersections; for example, it randomly chooses a direction to turn. Further, the students are able to use multiple smart educational robots as transmitters when interaction between them takes place. The robot can automatically react to receiving a certain signal from another robot and can then send a response. The third most common input/output interaction is to turn on a light based on the information received about the color of the surface the robot is currently on. These activities can also be used in lower grades to teach students. Such preprogrammed robot behavior can make the lesson more attractive, offering multiple possibilities even to students who do not yet know how to program, while showing them real-life applications of smart devices. For example, how a security sensor might react in the event of a theft.

7 Discussion and future research

In conclusion of our research, we consider it necessary to set the theoretical groundwork for continued research in this area of interactivity. Like any innovative technology, robotics has its limitations and challenges that need to be explored in detail. At the very least, the constantly changing market of available robot models and the rapid evolution of technology play a significant role in this research area. Even with ongoing research, there are still open questions for which we do not know the answers and which we plan to address in future research projects. For example, the previously mentioned motion is a large and important topic that belongs under output interaction because most smart educational robots can perform movement. However, they differ significantly in the way in which this motion is input to them, or rather in the programming environments and therefore in the commands. Since this would be a rather extensive result, we have chosen not to address it in this chapter.

When dealing with robotics in the context of education with a focus on programming, the type of sensor students work with is quite crucial. It depends on what type of data they are collecting and how challenging it is to work with that data in a programming language. Therefore, one rather big challenge in our research was and still is the handling of the different types of sensors, sound components, and lights that commonly available intelligent educational robots are equipped with. For this reason, it has been challenging to develop meaningful categories of robot interactivity, and at the same time, the question remains valid as to what extent it will be possible to formulate general statements or recommendations about the use of sensors in programming education in the future. This issue is complex and requires exploration of students' knowledge and competencies at different ages in order to determine the necessary predispositions for appropriate work with intelligent educational robots. Therefore, in future research, we plan to explore the type of interactivity (and specifically which sensors) that is appropriate to use for teaching beginners the basics of programming. We also plan to analyze learners' reactions to working with different sensors, sound components, and lighting elements during programming lessons. This will allow us to identify which sensors are more intuitive to pupils than others. Based on these particular findings, we intend to identify recommendations for types of interactivity that teachers should start with when teaching basic programming using intelligent educational robots. In a future research project, we would like to focus on students' perceptions, preconceptions, and their understanding of interactivity in educational robotics and also systematically analyze how teaching programming using educational robots and their sensors affects students' understanding of programming. Our goal will be to obtain a deeper understanding of how students construct their knowledge of programming using educational robotics. We believe that the results of this research project will contribute to a better understanding of the interactivity of intelligent educational robots and provide valuable insights that will help teachers and researchers better integrate robotics into education and facilitate the development of key competencies in students.

8 Conclusion

In the current era of digitalization and the rapid development of AI, it is essential that education systems reflect the new competencies and skills requirements of the next generation. We can conclude that the interactivity of smart educational robots brings not only a new dimension to the teaching of programming and algorithmic thinking but also increased motivation and active involvement of students in learning. It links theoretical knowledge to real life in a natural way, preparing students for their working lives and developing key skills needed for life in this century. Such a combination of teaching and real life is often a powerful tool in the hands of teachers to initiate students' inner motivation for active involvement in the learning process. Out-of-class life often provides students with very stimulating excitement and engaging activities that catch their attention, unlike regular teaching, which is often boring and unappealing to students. Therefore, we believe that there is a great need to pay research attention to educational robotics and other associated areas so that it can be integrated into education in a meaningful and appropriate way with respect to the cognitive adequacy of students. Therefore, the results in this chapter also attempt to contribute to a systematic view on this topic. The categorization of the sensors and output components that provide the interaction between the robot and the students (or other users) provides a suitable baseline for further exploration of appropriateness. With the categorization into input, output, and input-output interactivity, we can thus further explore to what extent working with input or output interactivity is challenging for learners. When, if ever, is it appropriate to include sensors in the teaching of programming with intelligent educational robots, and if so, which ones? Such findings will be able to significantly contribute to quality educational content or assessment of emerging curricula. Curricula on these topics are emerging, but there are no quality tools to assess them expertly.

The interactivity of smart educational robots, often combined with the use of AI, brings great potential for innovation in education but also raises concerns about content-appropriate activities and learning materials. Although it should have a significant place in the teaching process from an early age, more detailed research should be conducted into what is an appropriate way of introducing particular technologies, specifically sensors, into the educational process. We hope that this chapter has provided information that will help both in the creation of new, high-quality materials and in the selection of criteria by which experts can review these materials.

References

[1]   Dwivedi YK, Hughes L, Ismagilova E, Aarts G, Coombs C, Crick T, et al. Artificial Intelligence (AI): Multidisciplinary perspectives on emerging challenges, opportunities, and agenda for research, practice and policy. International Journal of Information Management, 2021 Apr, 57, 101994. doi: 10.1016/j.ijinfomgt.2019.08.002.

[2]   Tschang FT, Almirall E. Artificial Intelligence as augmenting automation: Implications for employment. Academy of Management Perspectives, 2021 Nov, 35(4), 642–659. doi: 10.5465/amp.2019.0062.

[3]   Chiu TK. A holistic approach to the design of artificial intelligence (AI) education for K-12 schools. TechTrends, 2021 Aug 12, 65(5), 796–807. doi: 10.1007/s11528-021-00637-1.

[4]   Xia L, Zhong B. A systematic review on teaching and learning robotics content knowledge in K-12. Computers & Education, 2018, 127, 267–282.

[5]   Sullivan FR. Robotics and science literacy: Thinking skills, science process skills and systems understanding. Journal of Research in Science Teaching: The Official Journal of the National Association for Research in Science Teaching, 2008, 45.3, 373–394.

[6]   Berry CA, Remy SL, Rogers TE. Robotics for all ages: A standard robotics curriculum for K-16. IEEE Robotics & Automation Magazine, 2016, 23(2), 40–46.

[7]   Karalekas G, Vologiannidis S, Kalomiros J. Europa: A case study for teaching sensors, data acquisition and robotics via a ROS-based educational robot. Sensors, 2020, 20.9, 2469.

[8]   Chakarov G, Alexandra, et al. Designing a middle school science curriculum that integrates computational thinking and sensor technology. Proceedings of the 50th ACM Technical Symposium on Computer Science Education. 2019.

[9]   Chung CJCJ. Integrated STEAM education through global robotics art festival (GRAF). In: 2014 IEEE Integrated STEM Education Conference. IEEE; 2014.

[10]   Fjukstad B, et al. Low-cost programmable air quality sensor kits in science education. Proceedings of the 49th ACM Technical Symposium on Computer Science Education. 2018.

[11]   Marcum-Dietrich N, et al. Our watershed: Students use data and models to make a difference in their own school yard. The Science Teacher, 2018, 85(2), 39–46.

[12]   Maloney JH, Peppler K, Kafai Y, Resnick M, Rusk N. Programming by choice. Proceedings of the 39th SIGCSE technical symposium on Computer science education. 2008 Mar 12, doi: 10.1145/1352135.1352260.

[13]   Eguchi A. Robocupjunior for promoting STEM education, 21st century skills, and technological advancement through robotics competition. Robotics and Autonomous Systems, 2016 Jan, 75, 692–699. doi: 10.1016/j.robot.2015.05.013.

[14]   Mikropoulos A, Bellou I. Educational robotics as mindtools. Themes in Science and Technology Education, 2013, 6(1), 5–14.

[15]   Oluk A, Korkmaz Ö, Oluk HA. Effect of scratch on 5th graders' algorithm development and computational thinking skills. Turkish Journal of Computer and Mathematics Education (TURCOMAT), 2018 Feb 28, doi: 10.16949/turkbilmat.399588.

[16]   Herschbach DR. The STEM initiative: Constraints and challenges. Journal of STEM Teacher Education, 2011, 48(1), doi: 10.30707/jste48.1herschbach.

[17]   Fischer K, Niebuhr O, Alm M. Robots for foreign language learning: Speaking style influences student performance. Frontiers in Robotics and AI, 2021 Sept 3, 8, doi: 10.3389/frobt.2021.680509.

[18]   Shahab M, Taheri A, Mokhtari M, Shariati A, Heidari R, Meghdari A, et al. Utilizing social virtual reality robot (V2R) for music education to children with high-functioning autism. Education and Information Technologies, 2021 Jan 2, 27(1), 819–843. doi: 10.1007/s10639-020-10392-0.

[19]   Ben Ari A. Nurturing computational thinking in an Israeli kindergarten with the cal-kibo robotics curriculum. AERA 2023, 2023, doi: 10.3102/ip.23.2012601.

[20]   Sterling L. Coding in the curriculum : FAD or foundational? 2016. (Accessed March 20, 2026, at https://research.acer.edu.au/research_conference/RC2016/9august/4)

[21]   Nouri J, Zhang L, Mannila L, Norén E. Development of computational thinking, digital competence and 21st century-skills when learning programming in K-9. Education Inquiry, 2019 Jun 13, 11(1), 1–17. doi: 10.1080/20004508.2019.1627844.

[22]   Resnick M. Sowing the seeds for a more creative society. Proceedings of the SIGCHI Conference on Human Factors in Computing Systems. 2009 Apr 4. doi: 10.1145/1518701.2167142.

[23]   Garneli V, Giannakos MN, Chorianopoulos K. Computing education in K-12 schools: A review of the literature. 2015 IEEE Global Engineering Education Conference (EDUCON). 2015 Mar. doi: 10.1109/educon.2015.7096023.

[24] Gibson S, Bradley P. A study of Northern Ireland key stage 2 pupils' perceptions of using the BBC micro:bit in stem education. The STeP Journal, 2017, 4(1), 15–41.

[25] Barker BS, Nugent G, Grandgenett N, Adamchuk VI. Robots in K-12 education: A new technology for learning. Hershey PA, USA: Information Science Reference; 2012.

[26] Socratous C, Ioannou A. Structured or unstructured educational robotics curriculum? A study of debugging in block-based programming. Educational Technology Research and Development, 2021, 69(6), 3081–3100.

[27] Merkouris A, Chorianopoulos K. Introducing computer programming to children through robotic and wearable devices. In: Proceedings of the Workshop in Primary and Secondary Computing Education. 2015 Nov 9. doi: 10.1145/2818314.2818342.

Ela Luria

Exploring the use of chatbots in English language learning among higher education students in Israel

Abstract: This study investigates the multifaceted impact of chatbot integration on language learning experiences among undergraduate students in Israel. Employing a mixed-method design, the study explores students' perceptions and interactions with chatbots in language learning, focusing on performance improvement, efficiency in learning, motivation sustainment, and self-confidence enhancement. A total of 100 undergraduate students undertaking English language courses participated in the study, providing both qualitative insights through semistructured interviews and quantitative data through a meticulously validated questionnaire. Thematic analysis of interview transcripts revealed recurrent themes related to the perceived effectiveness, convenience, engagement, and self-confidence building facilitated by chatbot interactions. Quantitative analyses demonstrated significant improvements across all dimensions of language learning outcomes following chatbot usage, supported by large effect sizes and positive correlations among various outcome measures. Mediation and regression analyses further elucidated the role of motivational and self-confidence factors in driving performance enhancement through chatbot interaction. Overall, the study provides compelling evidence of the beneficial effects of chatbot integration in optimizing language learning experiences, underscoring their potential as effective tools for enhancing language acquisition outcomes in higher education contexts.

Keywords: Chatbot, language learning, motivation, performance enhancement, second language acquisition

1 Introduction

In recent years, the integration of artificial intelligence (AI) technologies into education has gained considerable attention, offering innovative solutions to longstanding challenges faced by language learners worldwide [1]. With the rapid advancement of AI, particularly in the realm of natural language processing (NLP) and machine learning (ML), educational practitioners and researchers are increasingly exploring the potential of AI-driven tools to enhance foreign language learning experiences [2, 3].

Ela Luria, English and Education Department, Levinsky College of Education, Tel Aviv; Beit Berl College of Education, Beit Berl, Israel, e-mail: elaluria@yahoo.com

https://doi.org/10.1515/9783111352695-012

AI chatbots, in particular, have emerged as prominent agents of change in language education, leveraging sophisticated algorithms to simulate human-like interactions and provide personalized learning support [3]. These dynamic AI entities, capable of engaging learners in conversational interactions, hold the promise of addressing key challenges in language learning, including limited opportunities for practice and feedback, as well as the need for authentic communication experiences [4].

At the forefront of this transformation are pedagogical and conversational chatbots, which are designed to offer interactive engagements tailored to the needs and proficiency levels of individual learners [5]. Pedagogical chatbots, in particular, function as intelligent tutors, guiding learners through language exercises, providing feedback, and adapting instructional strategies based on learners' performance [6]. Conversely, conversational chatbots focus on facilitating natural language interactions, enabling learners to practice speaking, listening, and comprehension skills in a supportive virtual environment [7].

The effectiveness of AI chatbots in language education has been demonstrated across various educational contexts and language proficiency levels [8]. These chatbots leverage a range of communication modes, including text, speech, graphics, haptics, and gestures, to engage learners in meaningful language practice [9]. Moreover, their usability and accessibility, coupled with intuitive interfaces, make them well-suited for mobile platforms, allowing learners to access language learning resources anytime, anywhere [1, 10].

In addition to their instructional utility, AI chatbots have been shown to have a positive impact on learner motivation and engagement, key factors in language learning success [11, 12]. Text-based interactions, in particular, have been found to enhance motivation and reduce language anxiety, fostering a supportive learning environment conducive to language acquisition [13].

In summary, AI chatbots represent a transformative force in language education, offering learners personalized, interactive, and accessible opportunities for language practice and communication. As the educational landscape continues to evolve in the digital age, the integration of AI technologies holds the promise of revolutionizing language learning experiences and empowering learners to achieve proficiency and fluency in diverse linguistic contexts.

This chapter is structured to provide a comprehensive exploration of the impact of chatbot integration on language learning among higher education students in Israel. It begins with a literature review on the topic, outlining the current state of research and identifying gaps that this study aims to address. Following the literature review is a detailed description of the study's methodology, including the participants, data collection procedures, and the tools used for both qualitative and quantitative analyses. Section 2.4 presents an in-depth analysis of the qualitative data, highlighting themes such as perceived effectiveness, convenience, engagement, and self-confidence building. This is followed by the quantitative findings, which demonstrate significant improvements in language learning outcomes, supported by statistical analyses, including effect size calculations, correlation analysis, regression analysis, repeated measures ANOVA, and mediation analysis. The chapter concludes with a discussion that synthesizes the qualitative and quantitative results, emphasizing the overall benefits of chatbot usage in enhancing language learning experiences and outcomes.

1.1 Types and functions of AI chatbots

AI chatbots have emerged as versatile tools in language education, offering a range of functions and capabilities tailored to support language learners in various contexts [14]. This section explores the different types of AI chatbots and their respective functions in facilitating language learning experiences.

Pedagogical chatbots serve as intelligent tutors, guiding learners through structured language exercises and providing feedback based on their performance [15]. These chatbots are designed to scaffold learning activities, offering step-by-step guidance and personalized recommendations to help learners achieve their language learning goals. Through adaptive algorithms, pedagogical chatbots adjust their instructional strategies in real time, addressing learners' strengths and weaknesses to optimize learning outcomes [16].

In contrast to pedagogical chatbots, conversational chatbots prioritize natural language interactions, enabling learners to practice speaking, listening, and comprehension skills in authentic communicative contexts [17]. These chatbots simulate human-like conversations, engaging learners in dialogues and exchanges that mirror real-world communication scenarios. By providing opportunities for meaningful language practice, conversational chatbots enhance learners' fluency and communicative competence, fostering confidence and proficiency in the target language [8].

Some AI chatbots incorporate multimodal features, leveraging a combination of text, speech, graphics, haptics, and gestures to enhance the learning experience [9]. These chatbots offer diverse communication modes, catering to learners with different learning preferences and accessibility needs. By integrating multiple modalities, multimodal chatbots provide a rich and immersive learning environment, enabling learners to engage with language content in ways that are meaningful and engaging [1].

One of the key strengths of AI chatbots lies in their usability and accessibility, as they offer intuitive interfaces and flexible deployment options [12]. Chatbots can be accessed via various devices, including smartphones, tablets, and computers, allowing learners to engage with language content anytime, anywhere. Furthermore, chatbots with text-based interfaces are particularly well-suited for mobile platforms, providing seamless access to language learning resources on the go [13].

In summary, AI chatbots encompass a variety of types and functions, ranging from pedagogical tutors to conversational partners and multimodal interfaces. These chatbots play a crucial role in supporting language learners by providing personalized guidance, authentic communication opportunities, and flexible access to language resources.

1.2 Effectiveness of AI chatbots

AI chatbots have demonstrated significant promise in enhancing language learning experiences across diverse educational contexts. This section examines the effectiveness of AI chatbots in facilitating language acquisition and proficiency development among learners.

One of the key benefits of AI chatbots is their ability to improve students' communication skills in the target language [8]. By engaging learners in interactive dialogues and exchanges, chatbots provide opportunities for speaking, listening, and comprehension practice in authentic communicative contexts. Research has shown that regular interaction with chatbots leads to improvements in pronunciation, fluency, and communicative competence [17].

AI chatbots offer personalized learning support tailored to the individual needs and proficiency levels of learners [15]. Through adaptive algorithms, chatbots analyze learners' performance data and provide targeted feedback and recommendations for improvement. This personalized approach allows learners to progress at their own pace and focus on areas of weakness, thereby maximizing learning outcomes [16].

One of the strengths of AI chatbots is their ability to provide immediate feedback on learners' language production [14]. Unlike traditional classroom settings where feedback may be delayed or limited, chatbots offer real-time feedback on pronunciation, grammar, vocabulary usage, and other language skills. This immediate feedback enables learners to correct errors promptly and reinforce correct language usage, leading to more effective learning outcomes [1].

AI chatbots have been shown to enhance learner engagement and motivation through interactive and immersive learning experiences [12]. By simulating natural conversations and offering gamified language activities, chatbots capture learners' interest and maintain their attention throughout the learning process. Moreover, the novelty and interactivity of chatbot interactions contribute to a positive learning environment, fostering intrinsic motivation and enthusiasm for language learning [13].

AI chatbots offer accessibility and flexibility, allowing learners to access language learning resources anytime, anywhere [9]. Chatbots with text-based interfaces are particularly well-suited for mobile platforms, enabling learners to engage with language content on smartphones and tablets. This accessibility ensures that learners have convenient access to language learning support, regardless of their location or time constraints [7].

In summary, AI chatbots have emerged as effective tools for enhancing language learning experiences by improving communication skills, providing personalized learning support, offering immediate feedback, enhancing engagement and motivation, and promoting accessibility and flexibility.

1.3 Usability and accessibility

The usability and accessibility of AI chatbots play a crucial role in their effectiveness as tools for language education. This section explores how chatbots offer intuitive interfaces and flexible access to language learning resources, thereby enhancing the learning experience for diverse learners.

AI chatbots are designed with intuitive interfaces that facilitate user interaction and engagement [1]. Through user-friendly designs and NLP capabilities, chatbots create seamless communication experiences that mimic human-to-human interaction. Learners can easily navigate through chatbot interfaces, access language learning materials, and engage in interactive language practice activities without the need for extensive technical knowledge or training [8].

One of the key strengths of AI chatbots is their compatibility with mobile devices, allowing learners to access language learning resources on-the-go [13]. Chatbots with text-based interfaces are particularly well-suited for mobile platforms, as they require minimal bandwidth and can be accessed using smartphones and tablets. This mobile compatibility ensures that learners have convenient access to language learning support anytime, anywhere, thereby maximizing learning opportunities and flexibility [7].

AI chatbots offer adaptability to diverse learners with varying needs and preferences [14]. Chatbots can be customized to accommodate different learning styles, proficiency levels, and accessibility requirements. For example, chatbots can provide visual prompts for learners with hearing impairments, offer text-based interactions for learners with visual impairments, and adjust language difficulty levels based on learners' proficiency levels. This adaptability ensures that all learners have equitable access to language learning resources and support [15].

Some AI chatbots incorporate multimodal features, allowing learners to engage with language content using a combination of text, speech, graphics, and gestures [9]. By offering diverse communication modes, multimodal chatbots cater to learners with different learning preferences and accessibility needs. Learners can choose the mode of interaction that best suits their learning style and comfort level, thereby enhancing their engagement and participation in language learning activities [12].

AI chatbots help reduce barriers to access for learners from diverse linguistic and cultural backgrounds [17]. Chatbots can be programmed to support multiple languages, dialects, and cultural references, making them inclusive and welcoming to learners from different linguistic communities. Additionally, chatbots can provide language learning materials and support in remote or underserved areas where access to traditional language education resources may be limited.

In summary, AI chatbots offer intuitive interfaces, mobile compatibility, adaptability to diverse learners, multimodal interaction, and reduced barriers to access, thereby enhancing the usability and accessibility of language learning resources.

1.4 Motivation and engagement

Motivation and engagement are critical factors in language learning success, and AI chatbots have shown promise in enhancing these aspects of the learning experience. This section examines how chatbots stimulate learner motivation and engagement through interactive and immersive learning activities.

AI chatbots promote intrinsic motivation by offering engaging and interactive language learning activities [12]. Through gamified learning experiences, personalized feedback, and real-time interactions, chatbots capture learners' interest and maintain their enthusiasm for language learning. Learners are motivated to engage with chatbots voluntarily, driven by their own curiosity and desire to improve their language skills [13].

Chatbots provide learners with authentic communication experiences, which contribute to increased motivation and engagement [8]. By simulating natural conversations and real-world language use, chatbots enable learners to apply their language skills in meaningful contexts. Learners feel a sense of accomplishment and satisfaction when they successfully communicate with chatbots, which motivates them to continue practicing and improving their language proficiency [17].

AI chatbots offer immediate feedback and progress tracking features, which help learners stay engaged and motivated [1]. Learners receive instant feedback on their language production, allowing them to correct errors and monitor their progress in real time. This feedback loop reinforces learners' efforts and encourages them to persist in their language learning endeavors [16].

Chatbots tailor learning experiences to the individual needs and preferences of learners, enhancing motivation and engagement [14]. Through adaptive algorithms, chatbots analyze learners' performance data and adjust learning activities to match their proficiency levels, learning styles, and interests. Learners feel empowered and supported as they progress along personalized learning pathways, which fosters a sense of ownership and commitment to their language learning journey [15].

Some AI chatbots incorporate social interaction and collaboration features, which promote motivation and engagement among learners [9]. Learners can interact with peers, exchange language learning tips, and participate in collaborative learning activities facilitated by chatbots. This social dimension of chatbot interactions fosters a sense of community and camaraderie among learners, motivating them to actively participate and contribute to the learning process.

In summary, AI chatbots enhance motivation and engagement in language learning by promoting intrinsic motivation, providing authentic communication experiences, offering immediate feedback and progress tracking, tailoring learning pathways to individual learners, and facilitating social interaction and collaboration.

2 Method

2.1 Participants

A total of 100 undergraduate students from three universities in Israel were conveniently sampled for participation in this study. Participants were enrolled in English language courses at the Basic English Level. The gender distribution among participants was as follows: males (45) and females (55). The age distribution ranged from 18 to 29 years, with the majority falling within the 21- to 23-year category (45 participants).

2.2 Procedure

2.2.1 Qualitative data collection

Semistructured interviews were conducted with participants to explore their experiences with chatbots in language learning. Interviews focused on aspects such as performance expectancy, effort expectancy, motivation, and self-confidence. Participants were asked open-ended questions to elicit detailed responses regarding their interactions with chatbots and perceived benefits. Interviews were audio-recorded and transcribed verbatim for thematic analysis.

Thematic analysis of interview transcripts was conducted to identify recurring themes and patterns in participants' narratives. Researchers independently coded transcripts to identify relevant themes, which were then discussed and refined through iterative consensus-building sessions. Themes related to perceived effectiveness, convenience, engagement, and self-confidence building emerged from the qualitative data.

2.2.2 Quantitative data collection

Participants completed a meticulously validated questionnaire both before and after their interactions with chatbots. The questionnaire, developed specifically for this study, comprised Likert-scale items designed to assess participants' perceptions and experiences with chatbots in language learning contexts. The questionnaire items were informed by thematic insights derived from the qualitative analysis phase.

The same questionnaire was administered to participants before and after their engagement with chatbots to measure changes in perceptions and experiences over time. Participants were asked to rate their agreement with each statement on a scale ranging from 1 (strongly disagree) to 5 (strongly agree). The questionnaire items addressed dimensions such as performance improvement, efficiency in learning, motivation sustainment, and self-confidence enhancement.

The questionnaire underwent rigorous validation procedures to ensure its reliability and validity. Content validity was ensured through expert review, while pilot testing with a small sample of participants helped refine questionnaire items for clarity and relevance. Internal consistency reliability of the questionnaire items was assessed using statistical measures such as Cronbach's alpha coefficient.

2.3 Tools

In the quantitative phase of our research, we employed a meticulously validated questionnaire to investigate participants' perceptions and interactions regarding the utilization of chatbots in language learning contexts. The questionnaire underwent stringent validation procedures to ensure its appropriateness and efficacy in assessing multifaceted dimensions pertinent to language learning outcomes. Notably, the questionnaire's construction drew upon thematic insights derived from the qualitative analysis phase of our study.

The validation process involved several steps to establish the reliability and validity of the questionnaire:

Subject matter experts in language education and research methodology reviewed the questionnaire items to ensure they adequately covered the targeted dimensions, such as performance expectancy, effort expectancy, motivation, and self-confidence.

A pilot study was conducted with a small sample of participants to assess the clarity, comprehensibility, and relevance of the questionnaire items. Feedback from pilot participants was used to refine and improve the quality of the questionnaire.

Internal consistency reliability of the questionnaire items was assessed using statistical measures such as Cronbach's alpha coefficient. A high Cronbach's alpha value indicates strong internal consistency among the items, demonstrating the reliability of the questionnaire.

The questionnaire's ability to measure the intended constructs (e.g., performance improvement and efficiency in learning) was evaluated through factor analysis to confirm the underlying structure of the measurement model.

2.3.1 Questionnaire items

The questionnaire comprised Likert-scale items designed to capture participants' perceptions and experiences with chatbots in language learning. Participants were asked to rate their agreement with each statement on a scale ranging from 1 (strongly disagree) to 5 (strongly agree). The questionnaire items are presented in Tab. 1.

Tab. 1: Dimensions and average score.

Measure	Question about the method (traditional/chatbots) pre- and post-questionnaire
Performance improvement	This method (traditional/chatbots) has significantly improved my speaking skills in English.
	This method (traditional/chatbots) has enhanced my reading comprehension skills.
	This method (traditional/chatbots) has contributed to an improvement in my listening skills in English.
	I feel more confident in my writing proficiency in English.
Efficiency in learning	This method (traditional/chatbots) is time-efficient.
	This method (traditional/chatbots) has helped me achieve my language learning goals.
	This method (traditional/chatbots) has helped me with my overall efficiency in language learning.
	This method (traditional/chatbots) has streamlined and made my language learning process more effective.
Motivation sustainment	This method (traditional/chatbots) has helped maintain my motivation to practice English regularly.
	This method (traditional/chatbots) has helped me become more motivated to engage in language learning activities.
	My overall enthusiasm for language learning has increased since incorporating this language learning method (traditional/chatbots) into my study routine.
	I feel more encouraged to persist in overcoming language learning challenges after using this method (traditional/chatbots).
Self-confidence enhancement	After using this method (traditional/chatbots), I feel more confident in my ability to communicate effectively in English.
	This method (traditional/chatbots) has increased my self-assurance in engaging in real-life conversations in English.
	Using this method (traditional/chatbots) has positively impacted my confidence levels in expressing myself in written English.
	Overall, I feel more empowered and self-assured in my language abilities after consistent engagement with this method (traditional/chatbots).

2.4 Findings

2.4.1 Qualitative findings

The qualitative component of this study involved conducting semistructured interviews with participants to explore their experiences with chatbots in language learning, focusing on performance expectancy, effort expectancy, motivation, and self-confidence. Thematic analysis of interview transcripts was employed to identify recurring themes and patterns, allowing for a deeper understanding of participants' narratives.

2.4.1.1 Perceived effectiveness

Participants frequently discussed their perceptions of the effectiveness of chatbots in supporting their language learning journey. Many described how interacting with the chatbot improved their language skills, particularly in areas such as vocabulary acquisition, grammar comprehension, and pronunciation practice.

Quotes
1. "I found that using chatbots really helped me understand difficult grammar concepts."
2. "The chatbot helped me learn new vocabulary words in a more engaging way."
3. "I felt like my pronunciation improved after practicing with the chatbot."
4. "The chatbot provided instant feedback, which made it easier to correct my mistakes."
5. "I noticed a significant improvement in my reading comprehension skills."
6. "Chatbots made learning grammar less boring and more interactive."
7. "I liked how the chatbot adapted its responses based on my level of proficiency."
8. "The chatbot helped me reinforce what I learned in class through practice."
9. "Using chatbots made language learning feel more like a game than a chore."
10. "I felt more confident using English in real-life situations after interacting with the chatbot."

2.4.1.2 Convenience and accessibility

A prominent theme centered around the convenience and accessibility of chatbots as language learning tools. Participants appreciated the anytime, anywhere access to language practice provided by chatbots, highlighting the flexibility they offered in integrating language learning into their daily routines.

Quotes
1. "I could practice English anytime, anywhere with the chatbot."
2. "It was convenient to have a language learning tool on my phone."
3. "I liked being able to access the chatbot whenever I had free time."
4. "Chatbots made it easier to fit language practice into my busy schedule."
5. "The chatbot allowed me to learn at my own pace, which was really helpful."
6. "I appreciated the flexibility of being able to practice English online."
7. "Using chatbots eliminated the need to commute to language classes."
8. "It was convenient to have immediate access to language practice materials."
9. "Chatbots made language learning more accessible to people with busy lifestyles."
10. "I could practice English even when I was away from my computer."

2.4.1.3 Engagement and motivation

Participants often expressed how interacting with chatbots increased their motivation and engagement in language learning activities. They described feeling more motivated to practice and improve their language skills due to the interactive and engaging nature of chatbot interactions.

Quotes
1. "The chatbot made learning English more fun and engaging."
2. "I looked forward to practicing with the chatbot every day."
3. "The interactive nature of the chatbot kept me motivated to learn."
4. "I enjoyed challenging myself to improve my scores on the chatbot activities."
5. "The chatbot made me feel like I was progressing in my language skills."
6. "I felt motivated to practice English because the chatbot gave me instant feedback."
7. "Using chatbots made me more excited about learning English."
8. "I felt a sense of accomplishment every time I completed a chatbot lesson."
9. "The chatbot provided a sense of achievement when I mastered new language concepts."
10. "Chatbots made language learning feel like less of a chore and more like a hobby."

2.4.1.4 Self-confidence building

Many participants discussed how using chatbots boosted their self-confidence in their language abilities. They shared stories of feeling more confident in their speaking and writing skills after practicing with chatbots, attributing this increased confidence to the supportive and nonjudgmental nature of the interactions.

Quotes

1. "After practicing with the chatbot, I felt more confident speaking English with others."
2. "The chatbot helped me overcome my fear of making mistakes in English."
3. "I felt more comfortable writing in English after using the chatbot."
4. "Using chatbots improved my confidence in my language abilities."
5. "I gained confidence in my English pronunciation through repeated practice with the chatbot."
6. "The chatbot provided a supportive environment for me to build my language skills."
7. "I felt more self-assured in my English abilities after using the chatbot regularly."
8. "The chatbot encouraged me to take risks and try new language activities."
9. "I felt more confident in my ability to understand English conversations after using the chatbot."
10. "Chatbots helped me build my confidence by providing constructive feedback."

2.4.2 Quantitative findings

2.4.2.1 Performance improvement

Following the use of chatbots, participants demonstrated observable improvements in their language learning performance. Quantitative assessments revealed significant enhancements in various language skills, including speaking, reading, listening, and writing. Participants reported increased proficiency levels and greater fluency in English, as evidenced by their performance on language assessments and self-reported measures. Before using the chatbot, participants provided an average score of 3.2 on the questionnaire items related to performance improvement. After using the chatbot, this average score significantly increased to 4.5 on the Likert scale ($p < 0.001$). This demonstrates a substantial improvement in participants' perceptions of their language learning performance, including speaking, reading, listening, and writing skills.

2.4.2.2 Efficiency in learning

Participants experienced enhanced efficiency in their language learning endeavors as a result of engaging with chatbots. Quantitative analyses showed that participants were able to achieve their learning goals more effectively and efficiently with the support of chatbots. They reported spending less time on traditional language learning methods and more time engaging with chatbots, leading to greater productivity and progress in their language acquisition journey. Participants initially rated the efficiency of their language learning methods at an average score of 3.5 on the questionnaire items. Following the use of the chatbot, this average score rose to 4.3 on the Likert scale ($p < 0.01$). This indicates that participants perceived a marked increase in the efficiency of their language learning processes with the integration of chatbot technology.

2.4.2.3 Motivation sustainment

Chatbot usage contributed to the sustained motivation of participants in their language learning endeavors. Quantitative assessments revealed that participants maintained high levels of motivation and engagement throughout their interactions with chatbots. They reported feeling more motivated to practice English regularly, engage in language learning activities, and persist in overcoming challenges encountered during the learning process. Before using the chatbot, participants reported an average score of 3.8 on the questionnaire items related to motivation sustainment. After incorporating chatbots into their language learning routine, this average score increased to 4.6 on the Likert scale ($p < 0.001$). This suggests that chatbot usage contributed to maintaining high levels of motivation and engagement in language learning activities over time.

2.4.2.4 Self-confidence enhancement

Participants experienced a notable enhancement in their self-confidence levels as a result of using chatbots for language learning. Quantitative analyses demonstrated a significant increase in participants' confidence in their ability to communicate effectively in English, engage in real-life conversations, and tackle language-related tasks with confidence. They reported feeling more empowered and self-assured in their language abilities, leading to greater assertiveness and willingness to engage in language learning activities. Participants initially rated their self-confidence levels in language skills at an average score of 3.6 on the questionnaire items. After consistent engagement with chatbots, this average score rose to 4.4 on the Likert scale ($p < 0.05$) (see Tab. 2). This indicates a notable enhancement in participants' self-confidence levels in various language abilities, such as communication, conversation, and writing proficiency.

Tab. 2: Dimensions and average score.

Dimension	Mean score (Likert's scale)
Performance improvement	4.5
Efficiency in learning	4.3
Motivation sustainment	4.6
Self-confidence enhancement	4.4

2.4.2.5 Effect size (Cohen's *d*)

Cohen's d values were calculated to determine the effect size of chatbot usage on various dimensions of language learning outcomes, including performance improvement, efficiency in learning, motivation sustainment, and self-confidence enhancement. The effect sizes were interpreted based on Cohen's guidelines, where a value of 0.2 indicates a small effect, 0.5 indicates a medium effect, and 0.8 or higher indicates a large effect.

Performance improvement: The calculated Cohen's d for performance improvement was 1.50, indicating a large effect size. This suggests that the implementation of chatbots in language learning significantly contributed to enhancing participants' language skills, such as speaking, reading, listening, and writing proficiency. The substantial increase in performance underscores the efficacy of chatbots as effective tools for improving language learning outcomes.

Efficiency in learning: The Cohen's d value for efficiency in learning was determined to be 0.80, indicating a large effect size. This signifies a significant improvement in the efficiency of participants' language learning processes following the integration of chatbot technology. Participants reported spending less time on traditional methods and more time engaging with chatbots, resulting in greater productivity and progress in their language acquisition journey.

Motivation sustainment: The calculated Cohen's d for motivation sustainment was 1.20, indicating a large effect size. This highlights the substantial impact of chatbot usage on maintaining high levels of motivation and engagement among participants in their language learning endeavors. Participants reported feeling more motivated to practice English regularly and persist in overcoming challenges encountered during the learning process, reflecting the enduring positive influence of chatbots on motivation levels.

Self-confidence enhancement: The Cohen's d value for self-confidence enhancement was found to be 0.70, indicating a large effect size. This suggests a notable enhancement in participants' self-confidence levels across various language abilities, including communication, conversation, and writing proficiency, as a result of consistent engagement with chatbots. The increase in self-confidence underscores the empowering effect of chatbots on participants' language skills and confidence levels.

These effect sizes provide quantitative evidence of the substantial impact of chatbot usage on language learning outcomes, reinforcing the importance of integrating technology-driven approaches to enhance language acquisition experiences.

Correlation analysis revealed strong positive correlations between performance improvement and motivation sustainment ($r = 0.70$, $p < 0.001$) and between efficiency in learning and self-confidence enhancement ($r = 0.65$, $p < 0.001$). These findings suggest that improvements in one aspect of language learning are associated with improvements in other related aspects.

2.4.2.6 Regression analysis

Regression analysis indicated that efficiency in learning significantly predicted the performance improvement ($\beta = 0.45$, $p < 0.01$), suggesting that more efficient language learning methods lead to greater improvements in language skills. However, efficiency in learning was not a significant predictor of motivation sustainment or self-confidence enhancement.

2.4.2.7 Repeated measures ANOVA

Results of repeated measures ANOVA revealed a significant interaction effect between time (pre- vs. post-chatbot usage) and dimension of language learning outcomes ($F(3, 96) = 22.56$, $p < 0.001$). Post-hoc tests indicated significant improvements in all dimensions of language learning outcomes after chatbot usage compared to before, with performance improvement showing the largest increase ($F(1, 96) = 55.21$, $p < 0.001$).

2.4.2.8 Mediation analysis

Mediation analysis showed that the effect of chatbot usage on performance improvement was partially mediated by increased motivation (indirect effect = 0.32, $p < 0.001$) and enhanced self-confidence (indirect effect = 0.25, $p < 0.001$). These findings suggest that chatbot usage enhances performance improvement by increasing motivation and self-confidence in language learning.

In summary, the quantitative study findings revealed positive outcomes observed after participants used chatbots for language learning. These outcomes included improvements in language learning performance, enhanced efficiency in learning processes, sustained motivation levels, and increased self-confidence in language skills. These findings underscored the beneficial effects of chatbot usage on participants' language learning experiences and outcomes in the higher education context.

3 Discussion

The findings of our study underscore the positive impact of integrating chatbots into language learning environments. Across various dimensions of language learning outcomes, including performance improvement, efficiency in learning, motivation sustainment, and self-confidence enhancement, participants reported significant benefits following their interaction with chatbots.

3.1 Performance improvement

Participants demonstrated notable advancements in language learning performance across multiple skills such as speaking, reading, listening, and writing. This improvement was evidenced by both quantitative assessments and self-reported measures. The substantial increase in average scores from pre-chatbot to post-chatbot usage highlights the significant enhancement in participants' perceptions of their language proficiency [19]).

3.2 Efficiency in learning

The integration of chatbots facilitated more efficient language learning processes, as indicated by participants spending less time on traditional methods and more time engaging with chatbots [9]. This shift led to greater productivity and progress in their language acquisition journey. The substantial increase in the efficiency score further substantiates the perceived benefits of chatbot technology in optimizing language learning endeavors [7].

3.3 Motivation sustainment

Chatbot usage contributed significantly to sustaining high levels of motivation and engagement among participants. Their increased motivation to practice English regularly and persist in overcoming learning challenges underscores the enduring positive influence of chatbots on motivation levels [7]. The substantial rise in motivation scores emphasizes the role of chatbots in maintaining participants' enthusiasm for language learning activities over time [13].

3.4 Self-confidence enhancement

Participants experienced a notable enhancement in their self-confidence levels, particularly in communication, conversation, and writing proficiency, after consistent engagement with chatbots. This increase in self-assurance reflects the empowering effect of chatbots on participants' language skills and confidence levels. The significant rise in self-confidence scores underscores the transformative impact of chatbots on participants' perceptions of their language abilities [7].

3.5 Effect sizes

The calculated effect sizes (Cohen's d) further confirm the substantial impact of chatbot usage on language learning outcomes, with all dimensions exhibiting large effect sizes. These effect sizes provide quantitative evidence of the meaningful improvements observed in performance, efficiency, motivation, and self-confidence following chatbot interaction.

3.6 Correlation and regression analyses

The strong positive correlations between different dimensions of language learning outcomes suggest that improvements in one aspect of language learning are associated with improvements in other related aspects. Additionally, regression analysis revealed that efficiency in learning significantly predicted performance improvement, highlighting the role of efficient language learning methods in driving overall language skill enhancement.

3.7 Repeated measures ANOVA

The significant interaction effect between time (pre- vs. post-chatbot usage) and language learning outcome dimensions further supports the positive impact of chatbot usage on language learning. Post-hoc tests indicated significant improvements in all dimensions of language learning outcomes after chatbot usage, with performance improvement showing the largest increase.

3.8 Mediation analysis

Mediation analysis revealed that the effect of chatbot usage on performance improvement was partially mediated by increased motivation and enhanced self-confidence, underscoring the role of motivational and self-confidence factors in driving performance enhancement through chatbot interaction.

In conclusion, our study provides robust evidence of the beneficial effects of chatbot integration in language learning contexts. These findings highlight the potential of chatbots as effective tools for improving language learning performance, enhancing learning efficiency, sustaining motivation, and boosting self-confidence among learners. Such insights can inform the development and implementation of technology-driven approaches to optimize language learning experiences in educational settings

4 Conclusion

This study investigates the multifaceted impact of chatbot integration on language learning experiences among undergraduate students in Israel. Employing a mixed-method design, the study explores students' perceptions and interactions with chatbots in language learning, focusing on performance improvement, efficiency in learning, motivation sustainment, and self-confidence enhancement. Thematic analysis of interview transcripts revealed recurrent themes related to the perceived effectiveness, convenience, engagement, and self-confidence building facilitated by chatbot interactions. Quantitative analyses demonstrated significant improvements across all dimensions of language learning outcomes following chatbot usage, supported by large effect sizes and positive correlations among various outcome measures. Mediation and regression analyses further elucidated the role of motivational and self-confidence factors in driving performance enhancement through chatbot interaction.

The study provides compelling evidence of the beneficial effects of chatbot integration in optimizing language learning experiences, underscoring their potential as effective tools for enhancing language acquisition outcomes in higher education contexts. The implications of these findings suggest that educational institutions should consider incorporating AI-driven chatbots into their language learning curricula to provide personalized, engaging, and effective learning experiences. Future research should explore the long-term impacts of chatbot usage on language proficiency, investigate the effects of chatbots on different aspects of language learning such as cultural competence and pragmatic skills, and examine the potential of chatbots in other educational contexts and subjects. Additionally, further studies could focus on the development of more advanced chatbot features to cater to diverse learning needs and preferences, enhancing the overall effectiveness of AI-driven educational tools.

References

[1] Zhang S, Shan C, Lee JSY, Che S, Kim JH. Effect of chatbot-assisted language learning: A meta-analysis. Education and Information Technology, 2023, 28(11), 15223–15243.
[2] Yuan Y. An empirical study of the efficacy of AI chatbots for English as a foreign language learning in primary education. Interactive Learning Environments, 2023, 1, 1–16.
[3] Kohnke L. L2 learners' perceptions of a chatbot as a potential independent language learning tool. International Journal of Mobile Learning and Organisation, 2023, 17(1–2), 214–226.
[4] Jeon J, Lee S, Choi S. A systematic review of research on speech-recognition chatbots for language learning: Implications for future directions in the era of large language models. Interactive Learning Environments, 2023, 1, 1–19.
[5] Kohnke L. A pedagogical chatbot: A supplemental language learning tool. Regional Language Centre (RELC) Journal, 2023, 54(3), 828–838.
[6] Kuhail MA, Thomas J, Alramlawi S, Shah SJH, Thornquist E. Interacting with a chatbot-based advising system: Understanding the effect of chatbot personality and user gender on behavior. Informatics, 2022, 9(4), 81.
[7] Kim HS, Kim NY, Cha Y. Is it beneficial to use AI chatbots to improve learners' speaking performance?. Journal of Asia TEFL, 2021, 18(1), 161–178.
[8] Kuhail MA, Alturki N, Alramlawi S, Alhejori K. Interacting with educational chatbots: A systematic review. Education and Information Technology, 2023, 28(1), 973–1018.
[9] Gayed JM, Carlon MKJ, Oriola AM, Cross JS. Exploring an AI-based writing assistant's impact on English language learners. Computers and Education: Artificial Intelligence, 2022, 3, 100055.
[10] Ghafar ZN, Salh HF, Abdulrahim MA, Farxha SS, Arf SF, Rahim RI. The role of artificial intelligence technology on English language learning: A literature review. Canadian Journal of Language and Literature Studies, 2023, 3(2), 17–31.

[11] Junaidi J. Artificial intelligence in EFL context: Rising students' speaking performance with Lyra virtual assistance. International Journal of Advanced Science and Technology, 2020, 29(5), 6735–6741.

[12] Adam M, Wessel M, Benlian A. AI-based chatbots in customer service and their effects on user compliance. Electron Markets, 2021, 31(2), 427–445.

[13] Mukhallafi TRA. Using artificial intelligence for developing English language teaching/learning: An analytical study from university students' perspective. International Journal of English Linguistics, 2020, 10(6), 40.

[14] Fitria TN. Grammarly as AI-powered English writing assistant: Students' alternative for writing English. Metathesis, 2021, 5(1), 65–78.

[15] Mageira K, Pittou D, Papasalouros A, Kotis K, Zangogianni P, Daradoumis A. Educational AI chatbots for content and language integrated learning. Applied Science, 2022, 12, 3239. doi: 10.3390/app12073239.

[16] Sidorkin AM. Embracing chatbots in higher education: The use of artificial intelligence in teaching, administration, and scholarship. Taylor & Francis; 2024.

[17] Smutny P, Schreiberova P. Chatbots for learning: A review of educational chatbots for the Facebook Messenger. Computer Education, 2020, 151, 103862.

[18] Kohnke L. L2 learners' perceptions of a chatbot as a potential independent language learning tool. International Journal of Mobile Learning and Organisation, 2023, 17(1–2), 214–226.

[19] Yin Q, Satar M. English as a foreign language learner interactions with chatbots: Negotiation for meaning. International Online Journal of Education and Teaching, 2020, 7, 390–410.

[20] Rusmiyanto R, Huriati N, Fitriani N, Tyas NK, Rofi'i A, Sari MN. The role of artificial intelligence (AI) in developing English language learner's communication skills. Journal of Education, 2023, 6(1), 750–757.

[21] Kamalov F, Santandreu Calonge D, Gurrib I. New era of artificial intelligence in education: Towards a sustainable multifaceted revolution. Sustainability, 2023, 15(16), 12451.

[22] Silitonga LM, Hawanti S, Aziez F, Furqon M, Zain DSM, Anjarani S, Wu TT. The Impact of AI Chatbot-based Learning on Students' Motivation in English Writing Classroom. In: International conference on innovative technologies and learning. Cham: Springer Nature Switzerland; August 2023. 2023 pp. 542–549.

[23] Bakare OD, Jatto OV. The Potential Impact of Chatbots on Student Engagement and Learning Outcomes. In: Bonner E, Lege R, Frazier E, eds. Creative AI tools and ethical implications in teaching and learning. IGI Global; 2023. pp. 212–229.

[24] Zhai C, Wibowo S. A systematic review on artificial intelligence dialogue systems for enhancing English as foreign language students' interactional competence in the university. Computers and Education: Artificial Intelligence, 2023, 4, 100134.

[25] Azwary F, Indriani F, Nugrahadi DT. Question answering system berbasis artificial intelligence markup language sebagai media informasi. Klik-Kumpulan Jurnal Ilmu Komputer, 2016, 3(1), 48–60.

[26] Bahari A, Smith M, Scott H. Examining the impact of chatbot-based language learning support, adaptive learning algorithms, and virtual reality language immersion on EFL learners' language learning proficiency and self-regulated learning skills. 2024. This is a prepring available at: https://www.preprints.org/manuscript/202403.1715/v1

[27] Bonner E, Lege R, Frazier E. Large language model-based artificial intelligence in the language classroom: Practical ideas for teaching. Teaching English with Technology, 2023, 23(1), 23–41.

[28] Djalilova Z. Improving methodologies for integrative English and Latin language teaching using artificial intelligence technologies. Central Asian Journal of Education and Innovation, 2023, 2(12 Part 2), 29–34.

Svetlana Sharonova and Elena Avdeeva

Model of a teacher from the standpoint of ethics in education using artificial intelligence

Abstract: In this chapter, the ethical models of teachers of the twentieth century in the era of the fourth technological revolution are analyzed, and predictions for future models in the context of advanced artificial intelligence are provided. The basis for the reasoning was the ethical problems that had arisen for society as a result of the exploitation of actively developing artificial intelligence. The analysis highlighted specific ethical traits of teachers that artificial intelligence cannot simulate and that are key to societal development.

Keywords: Artificial intelligence, ethical teacher model, charisma, intuition, social imagination, identity

1 Introduction

With the development of artificial intelligence (AI), scientists are faced with ethical problems: ethical standards embedded in algorithms, ethical problems of human-machine (robot) interaction, and ethical problems of preserving the vital functions of robots. In 2005, Veruggio [1] was the first to declare the need to develop such a scientific direction as robotics. Of particular importance to ethical issues are the expectations for the next generation of robots, presented in the World Robot Declaration [2]:
(a) The next generation of robots will be coexisting partners with humans.
(b) The next generation of robots will help people both physically and psychologically.
(c) The next generation of robots will help build a safe and peaceful society.

However, technological development is happening so rapidly that already in 2015 the European Parliament [3] was forced to formulate the characteristics of autonomous (smart) robots: (1) the ability to become autonomous using sensors and (or) exchange data with their environment; (2) the ability to self-learn based on acquired experience; (3) availability of at least minimal physical support; (4) the ability to adapt one's actions and behavior in accordance with environmental conditions; and (5) absence of life from a biological point of view.

Corresponding author: Svetlana Sharonova, Department of Advertising and Business Communications, Economic Faculty, RUDN University, Moscow, Russia, e-mail: sharonova-sa@rudn.ru
Elena Avdeeva, JSC "Moscow Information Technologies", Russia, e-mail: el.v.avdeeva@gmail.com

https://doi.org/10.1515/9783111352695-013

The introduction of robots and AI into all areas has become an integral part of society. Education is no exception. As part of the first All-Russian Olympiad for school-children in AI in 2021, an online survey of finalists was conducted on the ethics of AI [4], and among students, there were responses with a proposal to replace teachers with AI.

The purpose of this chapter is to determine the teacher's model in the context of the integrated introduction of AI and robots into the education process. The original-ity of our approach lies in the fact that we will try to create a model of teacher behav-ior based on ethical problems that arise during the interaction between a teacher and an intelligent educational robot.

Initially, this chapter presents the main problems and issues within the field (Sec-tion 2). Thereafter, it presents the ethical models of teachers prior to the fourth techno-logical revolution (Section 3) and focuses on ethical AI models of teaching (Section 4). It further explores the ethical model of teachers in the era of AI (Section 5). Finally, con-clusive remarks and directions for future research are provided (Section 6).

2 Problem field

AI is defined as "the combination of cognitive automation, machine learning, reason-ing, hypothesis generation and analysis, natural language processing, and deliberate algorithm mutation to produce information and analytics at or above human capabili-ties" [5].

In education, robots and AI are being implemented in different forms: deep learn-ing, machine learning, social robots, 5G mobile networks, and intelligent systems. For all forms, there are fundamental tools: algorithms, large databases, sensors, and net-works. As practice has shown, a person also incorporates his/her inherent values and ethical standards into the algorithm. These values and ethical standards relate to the culture in which he/she raised and live. Modern technologies imbue us with the abil-ity to self-develop, achieving the creation of intelligence capable of solving problems on a par with the human mind. It would seem that the desire of students to replace teachers with AI is justified and quite feasible.

Using the example of social robots, one can see how AI reproduces emotive simu-lacrums of friendship, participation, and trust. However, when studying the ethical problems of students' communication with social robots, the importance of a human teacher is increasingly felt [6–10].

The teacher's model is considered from the perspective of competencies that are necessary in professional activities. However, the strong and integral side of the teacher's model is his/her values, ethics of behavior, intuition, and charisma. We will build a teacher model by contrasting these teacher qualities with AI simulacra. We will consider social robots as examples of physical embodiment and varieties of machine learning (tablets, computers, etc.) as virtual presence.

Social robots, according to Turkle [11], create a sense of attachment in people and create the illusion of human-to-human interaction [12], which may lead children to begin to prefer interacting with robots [13]. Virtual presence through tablets and computers also has a number of ethical problems: lack of emotive feedback, building a student's development trajectory on incorrect data due to the self-development of algorithms, and collecting personal data on students and teachers. The model of the teacher in education, when interacting with AI, assumes making the figure of the teacher central, and all technologies auxiliary for solving certain problems.

3 Ethical model of the teacher before the fourth technological revolution

As it is known, ethics, as a part of philosophy, deals with the identification of general principles of interpreting the phenomenon of morality. One of the trends in the development of ethics in the late twentieth century is the transition from a general philosophical discipline to applied ethics, which deals with moral collisions in particular spheres of social practice. The ethical model of a teacher reflects the moral and value norms of teacher-student interaction.

Based on the general principles of ethical standards, there are two models of interaction: **the teacher-centered model and the student-centered model** [14]. Throughout the twentieth century, these two models shaped and confronted each other. The teacher-centered model built a behaviorist framework that emphasized the teacher's didactic skills, official knowledge, and social spirit [15]. The idea of teacher competencies, which emerged in the 1960s, has become widespread since the 1980s. The goal of professional teacher education became specific measurable results: teachers' qualifications, scientific achievements expressed by the number of citations in international databases, the status of the university in international rating structures, and indicators of students' achievements in passing standardized tests [16].

The second model, the student-centered model, received multi-vector development. Relying on humanistic psychology and the theories of John Dewey and Donald Schön, the model took reflective activity on the part of the teacher as the basis for teacher-student interaction [15]. One such vector was sociocultural learning in a community of practice. Teachers' reflection on their own practice in a circle of like-minded people led to social, cognitive, and emotional development. Despite the fact that this reflection contributed to the formation of the teacher's professional identity [17], it also analyzed students' behavior and academic achievement. Thus, the teachers' focus was on the students, their situations, and ways to solve them.

Another aspect of the teacher-student model was Vygotsky's sociocultural and sociocognitive theoretical positions [18]. Here, in addition to the teacher and student, the surrounding circle of classmates is included. The process of interaction becomes more complex and multifaceted.

There are two views in relation to the transmission of culture in an educational setting:

– Free discourse of multiculturalism serves as a stimulus for cultural enrichment and the formation of international identity. This model is based on the principles of self-development and empowerment [19], focusing teachers' efforts not on the transfer of knowledge, but on the development of intellect, moral qualities, and aesthetics [20]. Meyers and Hermans' theory of the dialogical self [21] helps to understand the mechanisms of international identity formation, expanding the boundaries of the dialog between self and society due to the existence of multiple self-positions of perception and evaluation of certain phenomena. Thus, the teacher helps students to expand the boundaries of the dialog between self and society, demonstrating through their own experience the existence of multiple self-positions of perception and evaluation of certain phenomena.

– Reproduction of culture in the theories of Bourdieu [22] and Giddens [23]. The structure of the education system purposefully ensures the reproduction of existing elements of the social structure – classes. Giddens argued that structural values dominate the space of a particular structure (university); they are reproduced and created through the reflexive interaction between students and routine practices.

4 An ethical AI model of teaching without a teacher

4.1 Simulation of own experience

On the one hand, modern AI technology allows students to encounter other cultures while at home, thus expanding the boundaries of cultural interaction and gaining insights into the experiences of other cultures. On the other hand, it is the active use of gadgets that explains the lack of experience among young people. This is explained by the lack of real contacts with reality, which is filled by mediated experiences broadcast through the media, Internet, and social networks [24]. Freedom of choice has appeared for students, but the reference points in the form of traditional constants obtained in the course of reflection of personal experience acquire a fragmentary, elusive character. In search of stability and security, at least for a short period of time, students, provoked by modern technologies, tend to constantly acquire something and not only things but also feelings, affections, and perceptions [25]. The active integration of AI into education is leading to fundamental changes in ideas about the nature of knowledge. As Phillips [26] notes, the raw materials for formed concepts and understanding are memories of sensory impressions. AI uses neurological descriptions of these memories, relying on visualization and emotion [27]. People already get attached to the simplest robotic devices if they give them positive emotions and help them in their daily lives.

4.2 Replacement of the teacher

Ideally, modern technology can provide learning without a teacher. At the heart of such a model is machine learning. Machine learning involves computer programs that think like a human. Developing such programs involves creating an algorithm, which is a sequence of actions that should lead to a desired result. The algorithm does not answer the question posed; for this, it is necessary to explain what the expected output would be. The goal of machine learning is to teach the model to find a solution on its own. While searching for a solution, the machine itself identifies patterns using unlabeled data, bots, grouping them based on a set of attributes. Orientations to sound reasoning, situational awareness, and moral justice are shaped by the creator to avoid making bad decisions. However, as practice shows, the machine learns not only from positive but also from negative examples. Users tend to identify the robot's speech and situational behavior with human behavior [28]. People sometimes allow themselves to vent their stress in the form of verbal anger and aggression when communicating with bots and social robots. By gathering information from such negative examples, the machine learns negative vocabulary and aggressive behavior on its own and associates them with certain situations. Then it may begin to reproduce similar patterns in response. This shatters the expectations of the user, who is set up for impunity from the machine, the installation of which was meant to create and preserve peace of mind, safety, and comfort. Or, on the contrary, the machine supports and approves this behavior with its feedback.

Immersive technologies in education are of great importance in the era of the fourth revolution. The mastering of knowledge about objectively existing reality should be constantly accompanied by sensations because they play an important role in human life and activity, as they act as a source of knowledge about the world and about the person himself. Today, the creation of virtual environments allows virtual simulation of the three basic senses – sight, hearing, and touch.

AI helps to regulate student motivation issues in education through digital and computer games and simulations. Many studies acknowledge the effectiveness of using serious games. In addition to cognitive skills, games contribute to the development of social and emotional skills and enhance the ability to cooperate with peers [29, 30]. They provide moral and physical safety for future specialists, create a favorable emotional background in the classroom, strengthen the motivational component, offer an opportunity for multiple repetitions, and minimize psychophysical risk in critical nonstandard situations or during the first real practical experience [31].

4.3 Reproducing two traditional teaching models

In the era of the digitalization of education, elements of both the first and second ethical models described above are being used. The normative part providing control and management of education is built on the basis of measurement elements developed within the framework of the first teacher-centered model. Modern international structures and state management educational bodies are actively developing a list of competencies and multilevel educational standards. These criteria are the basis for monitoring the performance of educational institutions. As Raewyn Connell [16] points out, in such a model, there is no need for any conception of education as an intellectual discipline. The educational program is formed in accordance with the needs of the labor market, which is focused on meeting immediate needs. The market is not interested in long-term projects based on fundamental knowledge. According to Raewyn Connell [16], a competent teacher need not be able to reflect on the knowledge from which the educational program is derived. Under the new education management regime, the humanistic model of a good teacher is becoming an anachronism.

The goals of learning in the information space are changing. Now there is no need to memorize facts; the main thing is the ability to work with information. Since the student should be able to acquire knowledge on the Internet independently, the teacher's role is to mediate this process. In this case, it is not enough for students to have the necessary information; they must be active constructors of cognitive networks [32]. Analyzing learning interactions, understanding dialog, and analyzing multiple perspectives are all components drawn from the student-centered model for creating and maintaining networked relationships among members of learning groups. However, the practice of functioning of the second model in the conditions of digitalization of education deforms the expectations of students, as they are influenced by the requirements of official state bodies and have to worry about acquiring the knowledge necessary to pass final tests and examinations. Thus, the emphasis of new learning on developing flexible and decontextualized experiences, rather than memorizing facts and applying skills according to context, is almost negated [33].

5 The ethical model of the teacher in the AI era

As can be seen from the description of the model without a teacher, AI is able to recognize and simulate feelings and emotions, which are necessary for learning knowledge, calculating the educational trajectory of the student, and managing project work in game mode. So, what was left out of the picture? It is the lack of a real student experience. Experience enriches the ground for the development of creative intuition. Direct interaction with the teacher helps master such an experience.

The following traits of a good teacher's image, which cannot be simulated, are the key levers.

5.1 Teacher charisma

According to Irina Groza, "charisma is nothing more than an attractive force emanating from the personality; the art of mesmerizing other people" [34]. Psychologists studying the phenomenon of charisma identify a set of qualities that are characteristic of charismatic personalities. In particular, Shalaginova [35] lists 14 charismatic qualities. It would seem that if we take them as a basis, we can use a neural network to reproduce the charismatic image of a teacher. However, the power of the charismatic personality lies in her own satisfaction in the process, in the pleasure she gets in the here and now because she is doing what she truly enjoys. Therefore, a charismatic teacher is free-spirited, direct, and spontaneous, who is not afraid to be herself in all circumstances. Such spontaneity cannot be calculated by AI, and it is at the heart of shaping a student's personal experience.

5.2 Intuition

Intuition (from the Latin for "look into") refers to the capacity of knowing or understanding through direct insight, without rational analysis or deductive thinking. It can also refer to the mysterious psychological ability to obtain such knowledge [36]. However, such insights are not the result of the completely free, completely arbitrary activity of the human mind. It enters the sphere of interaction of the researcher's personality with the world of culture and information space. Thus, the richer the learner's personal experience, the more effective their intuition will be.

5.3 Goals

Another key aspect of the superiority of the human mind over the machine mind is the ability of humans to have more than one specific goal and their ability to update and change those goals. According to Andrei Zheleznov [37], AI outperforms humans in individual tasks, but the ability to perceive the overall context of different tasks, switch between them, and update goals remains exclusively human. Although AI was created precisely to act in situations for which it is impossible to prescribe a limited number of algorithms of action, the problem of morality got in the way of realizing this task. We cannot guarantee that an AI will always obey the moral constraints spelled out in instructions, since inherent in its idea is the ability to reinterpret them based on its own motivations and considerations. This perception of goals in terms of morality is the prerogative of only humans. Thus, the teacher, building learning objectives, is guided by a holistic picture, which includes not only disciplinary knowledge and skills but also moral implications in solving certain tasks and creating projects.

5.4 Social imagination

The affective sphere of relationships is associated with feelings and emotions. As we have already mentioned, AI is able to simulate and recognize the feelings and emotions of the communicator. However, the machine is not able to form an emotional attitude toward the phenomena of the surrounding world because it is not only an expression of feelings and emotions but also a person's reaction, reflecting his/her values, interests, and inclinations. A tool for developing students' awareness and processes of moral and social thinking is the social imagination. According to Taylor [38], social imagination is an unstructured understanding of what is going on in society and culture, an implicit grasp of social space by ordinary and common people that makes possible collaborative practices and a sense of moral order and legitimacy in relationships. According to Kristeva, "the imaginary is a kaleidoscope of my images, based on which the subject of the statement is formed" [39]. Social imagination allows a person to become a social being in order to realize oneself in everyday life and respond appropriately to life situations. In the learning process, these problems are addressed through aesthetic pedagogy, which offers teachers and students a sense of intellectual autonomy and can foster alternative ways of finding and creating meaning [40]. The aesthetic aspects of teaching are evident in the way a teacher positions himself/herself in relation to the subjects they teach. This allows for an exploration of the links between what the teacher knows about a subject and his/her personal response to that knowledge. The exclusion of live personal communication between teacher and student in conditions of teaching by AI transforms the nature of aesthetic understanding, which distorts the social adequacy of the student's emerging identity and personal positioning [41].

5.5 Identity of the teacher

Since education is involved in the creation of social reality, the figure of the teacher is key because the culture, morals, social environment, and goals with which the teacher identifies are the vectors for the development of society. The phenomenon of identity is the image of oneself acquired and accepted by an individual in interaction with the surrounding society. Erikson [42] considers identity as the internal continuity and identity of a person, formed in the process of development and performing adaptive functions. Although teacher identity is mobile, continuous, and changeable [43, 44], there are no fixed and established constructs from which to determine the factors of teacher identity.

Nevertheless, researchers have identified certain attributes [44–48]. The context for highlighting these attributes is not universal. Each set of teacher identity attributes reflects the sociocultural specificity of the academics' own perceptions. The presence of personality attribute descriptions is certainly used in AI developments. However, if teacher identification in live contact acts in a sense as an instrument of social control, in the case of AI, social control over the direction of development of society is either impossible to exercise due to the unpredictability of decisions and actions taken by AI, or, on the contrary, the issues of personal identification are under the strict control of the dominant group (network gurus).

6 Conclusion

In this chapter, the logic of reasoning is organized as follows:
- Initially, it was important to uncover the history of the emergence of interest in ethical issues within the use of AI in everyday life and technological developments.
- Then we tried to identify the problem field in the use of AI in educational environments, where we emphasized that the expanding capabilities of AI, which include knowledge of emotive sciences, allow us to hypothesize the replacement of a living teacher by a robot.
- The next step in our reasoning was to identify ethical models of teacher behavior in direct, live contact with students. In all the models described, there was a connection between historically developed social relations in society and attempts to theoretically explain these relations.
- After that, it was important to understand what ethical models of teacher behavior are considered by scientists who hold the position of replacing teachers with robots.
- Finally, it was interesting to find scientific concepts that do not contrast with the ethical model of a living teacher and a robot.

To summarize our considerations, we would like to focus not so much on the professional and digital competencies of teachers as on their personal characteristics. As we have tried to show, AI is able to replace the teacher in many routine practices, but it is not able to shape the lived experience of the young generation, to cause "a vital and personal experience" [49]. Internet space, along with modern media, simulates experiences in young people; at best, "we teach concepts instead of engaging our students in ideas" [50].

The personal characteristics of the teacher that we have identified: charisma, intuition, the ability to see generalized goals, to awaken the social imagination, and to identify oneself – all contribute to the unique ability to transmit their lived experience to students and to help them gain their own experience. The difference between such lived experiences in education and the simulated ones provided by AI is the transmission and capacity building of culture that contributes to the development of both students and teachers in progressive education. It is these qualities of teachers that will lead the way in the age of AI and allow humans to remain dominant over intelligent machines.

In conclusion, we would like to draw some prospects for further research in the stated problem field. Undoubtedly, AI will improve in bringing ever deeper emotive knowledge into the development of robot teachers, bots, and other forms of technologically created teacher images. The ethics of such technological forms of teacher behavior will develop along two (and perhaps in time more) lines: the ethics of treatment of techno-teachers and the ethics of techno-teacher-student interaction. There will certainly be a search for the most effective methods of learning in the digital space. A human will become increasingly aware of the need to maintain his/her advantage over AI. This will be expressed, on the one hand, in the awareness and expansion of routine tasks transferred to techno-pedagogues, and, on the other hand, in the emphasis on the formation of teacher and student personality on the development of special characteristics inherent in human consciousness and value orientations.

References

[1]   Veruggio G. The EURON Roboethics Roadmap. In: 6th IEEE-RAS international conference on humanoid robots. Geneva; 2006. pp. 612–617. doi: 10.1109/ICHR.2006.321337.

[2]   Veruggio G, Operto F. Chapter roboethics: A Bottom-up Interdisciplinary Discourse in the Field of Applied Ethics in Robotics. In: Machine ethics and robot ethics. London: Routledge; 2017. doi: 10.4324/9781003074991.

[3]   European Parliament. REPORT with recommendations to the commission on civil law rules on robotics (2015/2103 (INL)). Debates,15 February 2017. (Accessed February 2024 at: https://www.euro parl.europa.eu/doceo/document/A-8-2017-0005_EN.html)

[4]   Ryzhova NI, Trubina II. Updating the study of ethical problems of artificial intelligence by modern schoolchildren. Computer Science at School, 2022, 5(178), 26–31. doi: 10.32517/2221-1993-2022-21-5-26-31.

[5]   Saurabh K. Intelligent automation: The future. Journal of Robotics & Autom, 2021, 1(1), doi: 10.33552/ OJRAT.2021.01.000505.

[6]   Akgun S, Greenhow Ch. Artificial intelligence in education: Addressing ethical challenges in K-12 settings. AI Ethics, 2022, 2(3), 431–440. doi: 10.1007/s43681-021-00096-7.

[7]   Ghotbi N, Ho MT. Moral awareness of college students regarding artificial intelligence. Asian Bioethics Review, 2021, 13, 421–433. doi: 10.1007/s41649-021-00182-2.

[8]   Smakman M, Vogt P, Konijn EA. Moral considerations on social robots in education: A multi-stakeholder perspective computers & education. 2021, December, 174, 104317. doi: 10.1016/j. compedu.2021.104317.

[9]   Serholt S, Barendregt W, Vasalou A, Alves-Oliveira P, Jones A, Petisca S, Paiva A. The case of classroom robots: Teachers' deliberations on the ethical tensions. AI & Society, 2017, 32, 613–631. doi: 10.1007/s00146-016-0667-2.

[10] Holmes W, Porayska-Pomsta K, Holstein K, Sutherland E, Baker T, Buckingham Shum S, Santos OC, Rodrigo MT, Cukurova M, Bittencourt II, Koedinger KR. Ethics of AI in education: Towards a community-wide framework. Journal of Artificial Intelligence in Education, 2022, 32, 504–526. doi: 10.1007/s40593-021-00239-1.

[11] Turkle SA. Nascent robotics culture: New complicities for companionship. AAAI Technical Report Series. (Accessed February 2024 at: https://cdn.aaai.org/Workshops/2006/WS-06-09/WS06-09-010.pdf)

[12] Sharkey A, Sharkey N. Children, the elderly, and interactive robots. Robot Autom Mag IEEE, 2011, 18, 32–38. doi: 10.1109/MRA.2010.940151.

[13] Bryson JJ. Why robot nannies probably won't do much psychological damage. Interaction Studies, 2010, 11, 196–200. doi: 10.1075/is.11.2.03bry.

[14] Samuelowicz K, Bain JD. Conceptions of teaching held by academic teachers. Higher Education, 1992, 24, 93–111. doi: 10.1007/BF00138620.

[15] Arnon Sara N. Reichel who is the ideal teacher? Am I? similarity and difference in perception of students of education regarding the qualities of a good teacher and of their own qualities as teachers. Teachers and Teaching: Theory and Practice, 2007, 13(5), 441–464. doi: 10.1080/13540600701561653.

[16] Connell R. Good teachers on dangerous ground: Towards a new view of teacher quality and professionalism, Critical Studies in Education, 2009, 50(3), 213–229. doi: 10.1080/17508480902998421

[17] Wenger E. Communities of practice and social learning systems: The career of a concept. Social learning systems and communities of practice. 2010, 179–198. doi: 10.1007/978-1-84996-133-2.

[18] Vygotsky LS. Mind in society: The development of higher mental process. Cambridge, MA: Harvard University Press; 1978.

[19] Cohen-Almagor R. Just, reasonable multiculturalism: Liberalism, culture and coercion. New York and Cambridge: Cambridge University Press; 2021. doi: 10.1017/9781108567213.

[20] Mill JS. On Genius. In: John MR, Stillinger J, eds. The collected works of John Stuart Mill. volume I. autobiography and literary essays. Toronto: University of Toronto Press; London: Routledge and Kegan Paul 1981. doi: 10.1007/978-94-017-2010-6_3.

[21] Meijers F, Hermans H. The dialogical self -theory in education: A multicultural perspective. Cham: Springer International Publishing; 2018. doi: 10.1007/978-3-319-62861-5.

[22] Bourdieu P, Passeron J-C. Reproduction in education, Society and Culture. Sage Studies in Social and educational change 5. London: Sage Publications; 1977.

[23] Giddens A. The constitution of society: Outline of a theory of structuration. Cambridge: Polity Press; 1984.

[24] Yakushina OI. Identity in the Sociological Theory of E. Giddens. Modern Problems of Science and Education. 2014, 2.

[25] Entwistle N, Skinner D, Entwistle D, Orr S. Conceptions and beliefs about "good teaching": An integration of contrasting research areas. Higher Education Research & Development, 2000, 19(1), 5–26. doi: 10.1080/07294360050020444.

[26] Phillips DC. The good, the bad, and the ugly: The many faces of constructivism. Educational Researcher, 1995, 24(7), 5–12. doi: 10.3102/0013189X024007005.

[27] Wilson EO. Consilience. London: Little, Brown & Co; 1998.

[28] Abramov RN, Katechkina VM. Social aspects of human-robot interaction: Experience of experimental research. Journal of Sociology and Social Anthropology, 2022, xxV(2), 214–243. doi: 10.31119/jssa.2022.25.2.9.

[29] Shin S, Park JH, Kim JH. Effectiveness of patient simulation in nursing education: Meta-analysis. NurseEducation Today, 2015, 35(1), 176–182. doi: 10.1016/j.nedt.2014.09.009.

[30] Carenys J, Moya S. Digital game-based learning in accounting and business education. Accounting Education, 2016, 25(6), 598–651. doi: 10.1080/09639284.2016.1241951.

[31] Vaganova OI, Khokhlenkova LA, Voronina IR, Gushchin AV. Possibilities of simulation technologies in professorial education. Azimuth of Scientific Research: Pedagogy and Psychology, 2020, T. 9. № 3(32), 56–60. doi: 10.26140/anip-2020-0903-0010.

[32] Lunenberg M, Korthagen F, Swennen A. The teacher educator as a role model. Teaching and Teacher Education, 2007, 23, 586–601. doi: 10.1016/j.tate.2006.11.001.

[33] Anderson CW. Implementing Instructional Programs to Promote Meaningful, Self-regulated Learning. In: Brophy J, ed. Advances in research on teaching: Teaching for meaningful understanding and self-regulated learning. vol. 1. Greenwich, CT: JAI Press; 1989. pp. 311–343. https://doi.org/10.1007/978-94-011-4942-6_30

[34] Groza I. Charisma phenomenon as a matter of debates in the domestic and foreign psychology. National Association of Scientists (NAS), 2015, IV(9), 127–130.

[35] Shalaginova LV. Psychology of leadership. St Petersburg: Rech; 2007.

[36] New World Encyclopedia. (Available at: https://www.newworldencyclopedia.org/entry/Intuition)

[37] Zheleznov A. Morality for artificial intelligence: Prospects for philosophical rethinking. Logos, 2021, 31(6), 95–122. doi: 10.22394/0869-5377-2021-6-95-119.

[38] Taylor C. What is the social imaginary?. Emergency Reserve, 2010, 1(69), 19–26. doi: 10.1515/9780822385806-004.

[39] Kristeva J. A child with an unspoken meaning. Philosophical thought of France of the 20th century. Tomsk: Aquarius; 1998. pp. 297–305.

[40] Dina AM. Lutfi art and aesthetic education. Journal of Research in Philosophy and History, 2020, 3(1), 19. doi: 10.22158/jrph.v3n1p19.

[41] Hobbs L. Examining the aesthetic dimensions of teaching: Relationships between teacher knowledge, identity and passion. Teaching and Teacher Education, 2012, 28(5), 718–727. doi: 10.1016/j.tate.2012.01.010.

[42] Erickson E. Identity: Youth and crisis. Moscow: Flinta; 2006.

[43] Beijaard D, Meijer PC, Verloop N. Reconsidering research on teachers' professional identity. Teaching and Teacher Education, 2004, 20(2), 107–128. doi: 10.1016/j.tate.2003.07.001.

[44] Beauchamp C, Thomas L. Understanding teacher identity: An overview of issues in the literature and implications for teacher education. Cambridge Journal of Education, 2009, 39(2), 175–189. doi: 10/1080/03057640902902252.

[45] Izadinia M. A review of research on student teachers' professional identity. British Educational Research Journal, 2013, 39(4), 694–713. doi: 10.1080/01411926.2012.679614.

[46] Pennington MC, Richards JC. Teacher identity in language teaching: integrating personal, contextual, and professional factors. RELC Journal, 2016, 47(1), 5–23. doi: 10.1177/0033688216631219.

[47] Ahmad H, Latada F, Wahab MN, Shah SR, Khan K. Shaping professional identity through professional development: A retrospective study of TESOL Professionals. International Journal of English Linguistics, 2018, 8(6), 37. doi: 10.5539/ijel.v8n6p37.

[48] Kaplan A, Garner J, Semo. S. Teacher role-identity and motivation as a dynamic system. Chicago, IL: Paper presented at American Educational Research Association; 2015. doi: 10.13140/RG.2.1.1382.5684.

[49] Dewey J. The school and society and the child and the curriculum. Chicago: University of Chicago Press (Original work published 1902); 1990. doi: 10.7208/CHICAGO/9780226112114.001.0001.

[50] Kevin J, Pugh M. Girod science, art, and experience: constructing a science pedagogy from dewey's aesthetics. Journal of Science Teacher Education, 2007, 18, 9–27. doi: 10.1007/s10972-006-9029-0.

Chrissa Papasarantou, Dimitris Alimisis, and Elias Theodoropoulos

The AI-enhanced DIY robotic car: introducing the five big ideas of AI

Abstract: This chapter presents the "DIY robotic car that can be controlled using voice commands" project, developed within the framework of the AI4STEM Erasmus+ project. The project is a learning intervention designed to introduce teachers and students aged 12–16, to the fields of artificial intelligence (AI) and Internet of things (IoT) through the lens of the five big ideas, project-based learning pedagogies, and robotics. The project is divided into five activities. Each activity is accompanied by educational material, including teacher guidelines, student worksheets, and half-baked solutions, aiming to facilitate its implementation in the classroom. The project was piloted in a small-scale workshop by experienced STEM teachers and is currently being piloted with students from a junior high school, providing valuable insights into the effectiveness of using this approach for the introduction of AI and IoT in secondary education.

Keywords: Artificial intelligence, robotics, project-based learning, speech recognition, audio classification, Internet of things

1 Introduction

In recent years, many educational attempts have focused on finding ways to integrate artificial intelligence (AI) into the school curriculum [1–5]. The introduction of AI in education is identified as a key driver for innovative teaching and learning practices, facilitating, among other things, the digital transformation in education [6–9]. To this end, the acquisition of digital skills and competences, along with a number of twenty-first-century skills, including creativity, critical thinking, and problem-solving, is pertinent for both teachers and students [10]. One of the objectives identified by various

Acknowledgments: The authors express their gratitude to all the teachers who participated in the teachers' workshop and, particularly, Mr. George Fragkakis, the director of 23rd Gymnasium of Athens, who is piloting the project with his students.

Corresponding author: Chrissa Papasarantou, European Lab for Educational Technology EDUMOTIVA, Athens, Greece, e-mail: cpapasarantou@edumotiva.eu
Dimitris Alimisis European Lab for Educational Technology EDUMOTIVA, Greece, e-mail: alimisis@edumotiva.eu
Elias Theodoropoulos European Lab for Educational Technology EDUMOTIVA, Greece, e-mail: el.theodoropoulos@edumotiva.eu

https://doi.org/10.1515/9783111352695-014

researchers and organizations is the development of educational and training systems that can assist learners in gaining a comprehensive understanding of emerging digital technologies, including AI [10–13]. It is also of great importance to ensure that learners are adequately trained and prepared to utilize these technologies in a creative and beneficial, but also critical and ethical manner [8, 10]. Currently, there is a plethora of freely available AI tools and services [11, 14]. However, they are commonly utilized for the development of educational content, the support of teaching, as well as for student assessment [1, 5, 15]. Consequently, there is a lack of structured AI curricula that delve into the core concepts of AI [16], thereby assisting educators and learners in becoming familiar with the fundamental mechanisms of AI and ML through learning by doing in order to comprehend their functionality and build some relevant learning experience, thus avoiding the treatment of these concepts and services as black boxes [14, 17].

To this end, our previous research [18, 19] revolved around the creation of learning scenarios to introduce students to AI by combining this field with STEM practices, project-based learning [20–22], Maker education, and educational robotics [23, 24]. One of those research papers focused on presenting the learning scenario of a DIY robotic car that can be navigated through voice commands. This scenario was developed in the context of the EDU4AI project (2020–2022) (https://edu4ai.eu/), an Erasmus+ project aiming to showcase practices involving AI-based learning, coupled with methodologies that draw on the Maker Movement trend [25–27], and project-based learning, by using Arduino-based technology. The scenario was piloted by students aged between 15 and 17 years old, under the supervision of their teachers, leading to some fruitful findings concerning the implementation of such a learning intervention with respect to introducing AI in secondary education.

Drawing on this experience, this chapter presents an optimized solution for the DIY robotic car project that can be controlled and navigated through voice commands, developed in the context of the AI4STEM Erasmus+ project (https://ai4stem.erasmus plus.website/). The main goal of this new learning intervention is to familiarize both teachers and students (aged 12–16) with the five big ideas of AI as defined by the AI4K12 initiative (https://ai4k12.org/). This approach has been previously explored and adopted by other researchers as a means of facilitating AI learning [8, 10]. However, the present project seeks not only to introduce students to AI in light of the five big ideas but also to introduce them to the field of Internet of things (IoT). This will be achieved through the lens of project-based learning and robotics, and by using an appropriate microcontroller for educational purposes (i.e., BBC micro:bit), along with compatible electronic components and block-based programming environments. The project, which is divided into five main activities, is further described in the Section 3, followed by a number of learning objectives and learning prerequisites.

The project was presented and evaluated in a small-scale workshop by experienced STEM teachers, providing preliminary results on the usefulness of this learning intervention in STEM education, and specifically in familiarizing students with the field of AI. These findings, along with preliminary reactions of students from an ongoing piloting phase, are analyzed, leading to some conclusions and comments on the implementation of such a learning project as an introduction to AI and IoT in secondary education.

In brief, Section 2 provides a concise overview of the project entitled "Controlling a DIY Robotic Car with Voice commands," which was developed and implemented within the framework of the Edu4AI Erasmus+ project, and the main findings derived from the pilot testing of this project with students. Then, Section 3 presents the optimized solution of the project, which was developed in the context of the AI4STEM Erasmus+ project. This section is divided into five subsections. In particular, Section 3.1 identifies and presents the learning objectives that underpin this project, followed by Section 3.2, which determines the target group, the learning prerequisites, and the estimated time for implementing the project. Section 3.3 exhibits the activities that have been designed to facilitate the implementation of the project in the classroom setting. There are five different activities that are structured around the five big ideas, namely Perception (Section 3.3.1), Representation and Reasoning (Section 3.3.2), Learning (Section 3.3.3), Natural Interaction (Section 3.3.4), and Societal Impact (Section 3.3.5). Each paragraph provides detailed information regarding the methodology employed for the introduction of the five big ideas, along with the equipment/tools (hardware and software) utilized for this purpose. Section 3.4 then presents the findings of the demonstration and evaluation of the project in the workshop with STEM teachers, while Section 3.5 outlines the preliminary results of the pilots with students. These findings ultimately inform the conclusions presented in Section 4.

2 The Edu4AI implementation and findings

The project "Controlling a DIY robotic car with voice commands," developed in the Edu4AI Erasmus+ project, was about creating an Arduino-based robotic car with a certain degree of automation [18]. The main objectives of this project-based learning intervention were the construction and programming of the robotic car, as well as the design and programming of an application that would enable voice navigation through the integration of the speech recognition AI service. To this end, teachers' guidelines and students' worksheets, accompanied by videos and half-baked solutions, were produced to facilitate the smooth implementation of this project for both teachers and students. The project was piloted in two different countries (Greece and Spain) (https://www.youtube.com/watch?v=HZkyWBCu2cg) and was carried out by 37 students aged between 15 and 17 years, under the discreet supervision of four teachers. The students had no previous experience working with Arduino technology and had little programming experience. After the implementation of the project, both teachers and students completed a questionnaire to provide feedback and reflect on their experiences. Overall, this learning intervention was perceived as a successful method and an engaging learning process for familiarizing students with AI. The students stated that they liked this project and enjoyed the whole process of realizing it. In terms of difficulty, they stated that it was a project of average difficulty. Although the guidelines were clear, a lot of work had to be done to overcome difficulties and problems that arose, mainly in relation to the crafting and wiring process and the introduction to the IDE environment (https://www.arduino.cc/en/software). According to the teachers, a valuable experience gained during this project was the combination of crafting and programming to solve a problem in a real case scenario. Regarding AI, teachers reported that students realized how AI can be used in such projects. What they did not understand was the core mechanisms behind this field. This was also reflected in the students' feedback, who stated that they felt a bit confident in explaining what AI and ML are after the implementation of this project. In the teachers' opinion, longer introductory courses on AI would help students in this direction, increasing their confidence in implementing AI in future projects.

3 The AI4STEM approach

Drawing on this experience, an optimized solution for the DIY robotic car project was developed within the framework of the AI4STEM Erasmus+ project. The project focuses on the creation of a DIY robotic car that can be controlled using voice commands while collecting various data, including temperature, light level, distance, and acceleration, leading to certain decisions to optimize its performance. To simplify the wiring process, the BBC micro:bit microcontroller was chosen instead of Arduino. The BBC micro:bit is a microcontroller designed for educational purposes, making it a user-friendly and appealing option for students [28]. Furthermore, it is supported by a block-based programming environment, which simplifies the programming process [29].

This project-based learning intervention aims to introduce both teachers and students to the fields of AI and IoT, through the lens of the five big ideas (namely Perception, Representation and Reasoning, Learning, Natural Interaction, and Societal Impact), as suggested by the creators of the AI4K12 initiative. The project focuses on robotics and includes hands-on and computer-based practices and activities. Regarding AI, teachers and students are introduced to speech recognition and how this service can be used to navigate a robotic artifact. To this end, they are initially introduced to the processes of designing and programming an AI-based application using the MIT App Inventor software. Next, they are introduced to machine learning (ML) by using an ML tool (Personal Audio Classifier; https://c1.appinventor.mit.edu/) to train a model for classifying incoming navigational commands. Concerning robotics, they learn how to build a robotic car using the BBC micro:bit microcontroller and compatible electronic components, as well as how to program this robotic artifact using Microsoft Makecode block-based software (https://makecode.microbit.org/). To facilitate the implementation of this extensive learning intervention, the project is divided into five activities. Each activity revolves around one of the five big ideas. During these activities, students will focus on different parts of the implementation process and engage with different aspects of AI and IoT. The activities include guidelines for teachers and several suggested tasks for students to introduce and implement the concepts and develop twenty-first-century skills like creativity, critical thinking, problem-solving, and collaboration.

3.1 Learning objectives

After the completion of this project, and with respect to AI, the learners should be able to understand how AI services such as speech recognition work, explain basic programming concepts related to the implementation of speech-to-text methods, and identify the advantages, disadvantages, and risks of using voice commands in daily activities such as driving. With respect to robotics, the learners should be able to identify, discuss, and understand the role of AI in robotics, as well as explain and discuss different aspects of integrating AI into a robotic project (especially through speech recognition and voice commands). Concerning data structures and algorithms, they should be able to understand how decision trees or flowcharts can show possible logical paths and lead to decisions about operators that can be used at a later stage in programming. Regarding ML, they should be able to discuss the role of ML in AI robotics and how ML can be used to train a robotic artifact to perceive its surroundings. Especially with respect to audio classification, they should be able to explain the underlying basic concepts, identify and discuss the advantages and risks of audio classification, and understand that services such as speech recognition are prone to error if the trained model is biased. Finally, with respect to IoT, they will be able to monitor data in real time using an IoT analytics platform service and think about the impact of making data-driven decisions in everyday life.

The learners will also learn to construct a robotic artifact, create circuits as part of a robotic construction, create decision trees and flowcharts to represent a type of information, use programming commands coupled with AI methods to address specific behavior to a robotic artifact, program a robot to be instructed using voice commands, use the Personal Audio Classifier ML tool to classify different sounds, evaluate the results produced by an ML tool, make improvements to a trained model based on evaluation, and investigate factors in the training data that may lead to biases.

3.2 Target group, learning prerequisites, and implementation time

The project is aimed at students between the ages of 12 and 16 years old. Students should have some experience with block-based programming environments. It is a rather extensive project, needing several hours to properly address all the included aspects, while the duration may vary depending on the age and level of the students. For more experienced students, the time of implementation is estimated to be 15 h. For less experienced students, 30 h are foreseen. To facilitate the implementation process, especially for younger and less experienced students, a number of half-baked coding files are provided.

3.3 The activities of the project

3.3.1 Activity 1: introducing the big idea of perception through IoT

The first activity is a warm-up activity to introduce students to the big idea of Perception (https://ai4k12.org/big-idea-1-overview/) in the light of IoT and Robotics. Students explore how their robotic car can sense and perceive its environment by collecting and storing data that can later be used to make some decisions about the performance of the car. In particular, they explore how sensors built into the BBC micro:bit board (such as temperature and light sensors) or sensors that can be connected to the board (such as an ultrasonic sensor) can collect data and be monitored using an IoT analytics platform service. The service used for the needs of this activity is ThingSpeak (https://thingspeak.com/), a platform that allows aggregation, visualization, and analysis of live data streams in the cloud, while the ESP8266 Wi-Fi module is connected to

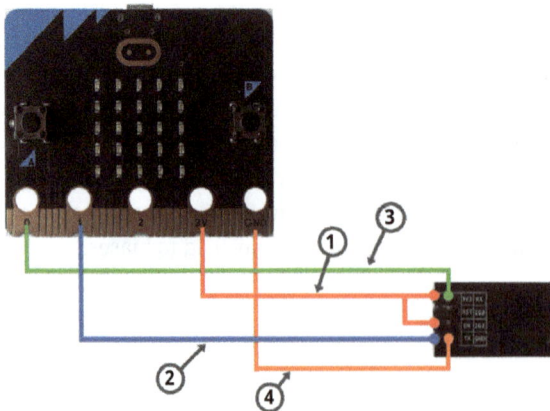

Fig. 1: Diagram for connecting the ESP8266 WiFi module to the BBC micro:bit microcontroller.

the micro:bit to enable the transfer of the collected data to this platform. The students create and program the circuit illustrated in Fig. 1, and then they observe the data monitored in the ThingSpeak platform. Based on their observations, they discuss how such data can be used to optimize the performance of the robotic car. The estimated duration of this activity varies from 4 to 6 h.

3.3.2 Activity 2: introducing the idea of representation and reasoning

The second activity aims to introduce students to the big idea of Representation and Reasoning (https://ai4k12.org/big-idea-2-overview/) and to familiarize them with the way in which a robotic artifact can "think" and "learn" from data. Students learn how data structures (namely constructing representations of the world used by intelligent agents) can be represented, and how algorithms can be used to extract specific information from these representations. To this end, they are initially encouraged to create some decision trees to understand the criteria for selecting an algorithm that leads to the best possible solution. Then, they create a decision tree (Fig. 2) to represent the decision that their robotic car should make when a voice command is received. Through this decision tree, they decide which logical path to follow when programming an application that allows the user to navigate the robotic car using voice commands. After that, they learn how to design and program this application by using the MIT App Inventor software (https://appinventor.mit.edu/) and the Speech Recognition AI service. The application (designed to be installed in a smart device) includes a button ("Press to Speak" in Fig. 3) that enables the smart device to record voice commands and use them as navigation

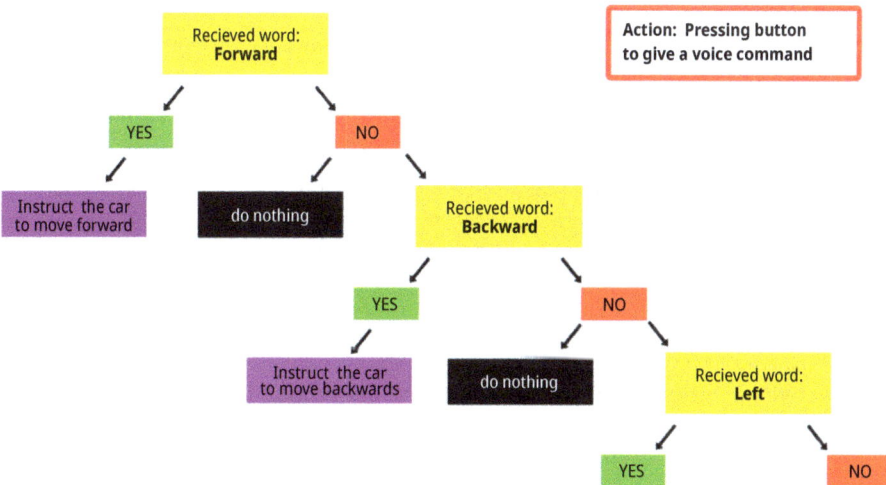

Fig. 2: Indicative decision tree, depicting the possible actions that can be activated when the "Press to Speak" button is pressed in the designed application.

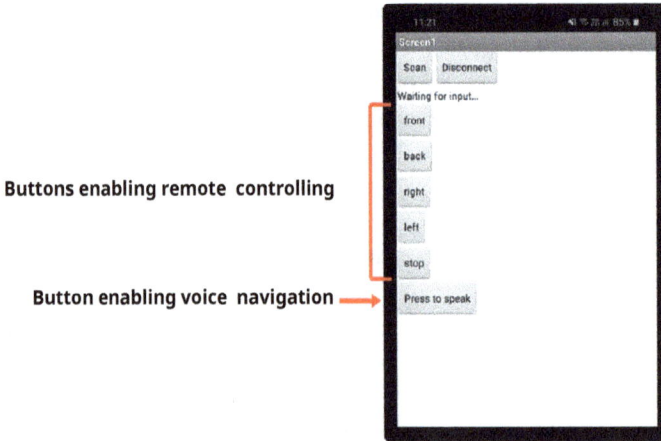

Fig. 3: Preview of the designed application.

commands for the robotic car. The estimated duration of this activity is between 8 and 15 h. This timeframe also includes a number of warm-up and preparatory activities, such as the creation (construction and wiring) and the programming of the robotic car, and the creation of an application that enables the remote controlling of the robotic car (Fig. 3).

3.3.3 Activity 3: introducing the idea of learning by training a model for recognizing voice commands

The third activity introduces students to the big idea of Learning (https://ai4k12.org/big-idea-3-overview/) by training a model to recognize specific voice commands. Through this activity, students understand the role of ML and ML algorithms in assisting a robotic artifact (in this case, the robotic car) to learn. In particular, they learn how to use the Personal Audio Classifier ML tool to train, test, evaluate, and export a model that classifies voice commands according to specific criteria (Fig. 4) and can be integrated (at a later stage) into the DIY robotic car. The estimated duration of this activity varies from 2 to 4 h.

3.3.4 Activity 4: introducing the idea of natural interaction by integrating a trained model into an AI application

In the fourth activity, the students are introduced to the big idea of Natural Interaction (https://ai4k12.org/big-idea-4-natural-interaction/). Specifically, they learn how to integrate the trained model from the third activity into the application created in

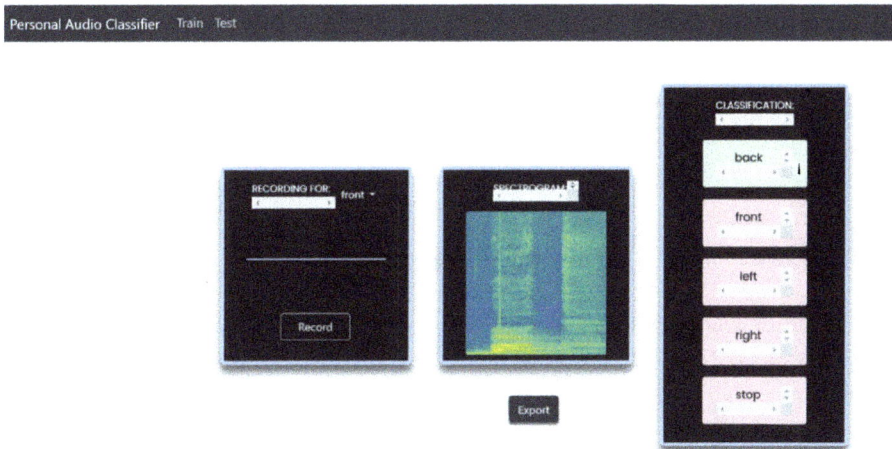

Fig. 4: An indicative example of the testing phase of a model trained to recognize and classify the following voice commands: "back," "front," "left," "right," and "stop." The trained model identified the "back" voice command (highlighted in green) in the recorded audio.

the second activity to observe how it affects the performance of the robotic car. To achieve this, the application produced in the second activity is reprogrammed to include a record button that enables the classification of the recorded voice command (Fig. 5). Based on the result of the classification, the robotic car is instructed to move accordingly. For example, if the recorded command is "front," the robotic car will

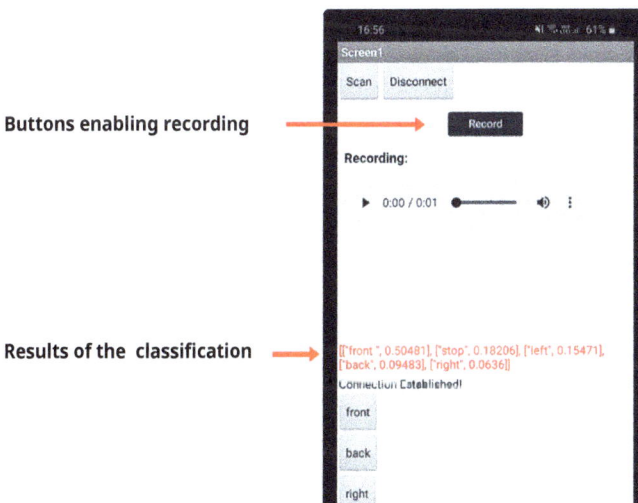

Fig. 5: Preview of the application after integrating the trained model, produced by the Personal Audio Classifier. In this example, the recorded voice command was recognized as "front" .

move forward. However, when the trained model is integrated into the application, it does not always function correctly, leading to malfunctions. In this way, this activity aims to raise students' awareness of the limitations of AI systems in interacting in a natural way, leading to errors. The estimated duration of this activity is between 2 and 4 h.

3.3.5 Activity 5: introducing the idea of societal impact

This activity introduces students to the big idea of Societal Impact (https://ai4k12.org/big-idea-5-societal-impact/) by reflecting on the experiences they have gained while carrying out the previous four activities. In particular, they are encouraged to think about the advantages, disadvantages, and inherent risks of using AI services and tools, as well as monitoring data and making decisions based on specific datasets. This activity can be implemented separately or blended with the previous four. In this way, the students become aware of several ethical decisions that need to be taken into account when designing and using AI and IoT services and technologies. The estimated duration of this activity is between 1 and 2 h.

All these activities are accompanied by educational resources (such as teachers' guidelines, students' worksheets, half-baked solutions, and circuit maps) that can help both teachers and students carry out the project smoothly. Through this project, it is expected that students will become more confident in using all the aforementioned methods and tools, and this entire learning experience will help them develop several twenty-first-century skills such as creativity, critical thinking, problem-solving, and collaboration.

3.4 Demonstrating the project with teachers

To test these hypotheses, experienced STEM teachers evaluated this project in a small-scale workshop. Eleven teachers (two of whom had participated in the pilots of the Edu4AI project) from primary, secondary, and vocational schools attended the two-hour workshop where they learned about the project's content and all the proposed activities and steps for implementation. They also interacted with the robotic artifact, examined the code, and shared their ideas and thoughts. Following the workshop, the teachers (10 out of 11) completed an online questionnaire (https://forms.gle/Lxc5miQg4nXy6FGP6) consisting of 14 questions for evaluating different aspects of the project. The questionnaire provided preliminary results on the effectiveness of this project in STEM education and in introducing students to the field of AI and IoT.

Seven questions focused on comprehension and how various project aspects can aid students in understanding AI and IoT (Tab. 1). Overall, the majority of teachers expressed positivity toward the effectiveness of this project in the comprehension of AI

Tab. 1: Teachers' feedback on how the project can aid students in understanding AI and IoT (from strongly disagree = 1 to strongly agree = 5).

The project can help students understand:	Strongly disagree	Disagree	Neutral	Agree	Strongly agree	Mean average
Issues related to the application of AI in everyday life	0	0	0	2	8	4,8
The Internet of things (IoT)	0	0	1	3	6	4,5
The first big idea (Perception)	0	0	0	1	9	4,9
The second big idea (Representation and Reasoning)	0	0	0	4	6	4,6
The third big idea (Learning)	0	0	0	3	7	4,7
The fourth big idea (Natural Interaction)	0	0	0	3	7	4,7
The fifth big idea (Societal Impact)	0	0	0	5	5	4,5

and IoT. Specifically, when asked whether the project can help students understand the application of AI in everyday life, eight teachers strongly agreed and two agreed. In terms of understanding IoT, six teachers strongly agreed, three agreed, and one was neutral. For the first big idea, nine strongly agreed and one agreed. For the second big idea, six strongly agreed and four agreed. As for the third and fourth big ideas, seven strongly agreed and three agreed, while for the fifth big idea the responses were equally divided between "strongly agree" and "agree."

Two questions focused on the project's concept (which involves teaching AI using a robotic artifact) and its potential implementation in the classroom (Tab. 2). Again, the majority of teachers expressed positivity toward these two items. Specifically, eight teachers strongly agreed and two agreed with the statement, "From a learning point of view, I find it useful to teach AI using a robotic artefact." When asked if they would implement this project with their students, five teachers strongly agreed and five agreed.

Tab. 2: Teachers' feedback on the project in terms of teaching and implementation (from strongly disagree = 1 to strongly agree = 5).

	Strongly disagree	Disagree	Neutral	Agree	Strongly agree	Mean average
From a learning point of view, I find it useful to teach AI using a robotic artifact	0	0	0	2	8	4,8
I would implement this project with my students	0	0	0	5	5	4,5

In addition, teachers were asked to rate the hardware (BBC micro:bit instead of Arduino) and software (Makecode and App Inventor block-based environments) on a scale ranging from very negative to very positive, and assess how much the proposed technology simplifies the project and facilitates its implementation for students (Tab. 3). Overall, the teachers were very positive (9 out of 10) about both the use of the micro: bit and the block-based programming environments to implement the project. This was also reflected in the discussion with them during the demonstration of the project and the interaction with the robotic artifact. However, concerns were raised about the applicability of the App Inventor to younger students (under 13 years old).

Tab. 3: Teachers' assessment of the hardware (BBC micro:bit) and the software (MakeCode and App Inventor).

	Very negative	Negative	Neutral	Positive	Very positive
In terms of simplifying the project and making it easier for students to implement, how do you assess the use of the micro:bit board instead of the Arduino?	0	0	0	1	9
How do you assess the fact that the project is implemented using only block-based code?	0	0	0	1	9

Teachers were asked to rate the ease of implementing the project using the materials provided, including student worksheets, half-baked solutions, and circuit diagrams, with discreet guidance. Nine teachers stated that it would be rather easy to implement the project, while one did not express an opinion. The teachers were also asked to suggest improvements that could help in the implementation of the project. Only three teachers provided written recommendations. Two of them suggested providing more half-baked or even complete solutions to facilitate implementation with younger students (under 13). Another teacher proposed extending the scenario to include the creation of a wheelchair navigated through voice commands, aligning with two of the 17 sustainable development goals set by the United Nations (https://sdgs.un.org/goals). Finally, the teachers were asked to estimate the number of teaching hours required to implement this project in school. Responses varied from 4 to 15 h, with a mean average closer to 10 h.

The results of the questionnaires were positive, supporting the proposed project as a suitable learning scenario for the introduction of AI and IoT in secondary education through the lens of the five big ideas. The discussion with the teachers revealed further interesting findings. According to them, the first three activities of the project are likely to run smoothly and excite students. Some expressed concerns about the fourth activity, as the trained model is prone to errors when integrated into the application, which could discourage students from completing the project. However, others expressed the opinion that this is a necessary step to make students realize that AI is not flawless and that this will increase their awareness of the disadvantages and risks associated with the use of AI. Despite these concerns, the majority of them were highly interested in having access to the educational material and the opportunity to implement the project with their students.

3.5 Piloting the project with students and preliminary results

The project is currently being piloted by a team of 10 junior high school students (13–15 years old) at the 23rd Gymnasium of Athens, under the supervision of an IT teacher (Fig. 6). The team meets once a week for 1 or 2 h after school. They have had eight meetings so far and are now working on the second activity of the project.

Based on the teacher's feedback, the students have enjoyed the process so far. However, they required additional time for the warm-up activities, particularly the younger students who lacked prior experience with a microcontroller, including the micro:bit board. They only had some programming experience in block-based environments such as Scratch (https://scratch.mit.edu/). Therefore, they dedicated some meetings to introductory activities such as programming the micro:bit LED screen to display messages, turning on and off an external LED light, and using the LED display as a light sensor. They did not encounter any difficulties in using the micro:bit and enjoyed this introductory process. After that, they proceeded with the creation of the robotic car, beginning with the construction and then moving on to the wiring. To facilitate the process, a pre-made cardboard chassis and a set of screws, spacers, and nuts were provided. They really enjoyed the construction of the robotic car, despite some difficulties in properly binding all the components to the chassis. Warm-up activities were then carried out in Makecode to program and test the functionality of the robotic car. They were very excited when they successfully programmed the robotic car to execute a variety of movements such as forward and backward motion. Meanwhile, the teacher engaged the students in a discussion about AI and its presence in daily life. The students mentioned applications such as ChatGPT and Bixby and expressed their eagerness to explore how they can create their own AI application.

Overall, based on the experience so far, the teacher finds the project very interesting and believes it offers opportunities to introduce several aspects of AI through a hands-on approach. He stated that the provided material, including teacher guidelines, student worksheets, and files with half-baked and complete programming solutions, are very helpful for implementing the project. However, he had to make adjustments to the implementation steps. For instance, he decided to temporarily skip the first activity and the aspect of IoT and focus solely on the AI aspect. This was done to align the needs of the project with the students' level, preventing them from becom-

Fig. 6: Images from the pilot. Left: a team examines the equipment; right: a team is programming the robotic car using the Microsoft MakeCode environment.

ing disappointed or discouraged. The pilots will continue for an additional month, during which both the teacher and his students will provide further feedback.

4 Conclusion and future plans

This chapter presents the project-based learning intervention "Controlling a DIY Robotic Car with Voice Commands," as developed within the framework of the AI4STEM Erasmus+ project. The project aims to introduce students aged 12–16 to the field of AI and IoT through robotics, focusing on the five big ideas. Specifically, through the five activities and associated tasks that revolve around the construction and programming of the robotic car, the design and programming of the AI application, as well as the use of an ML tool (i.e., Personal Audio Classifier), the students will gain a better understanding of the five big ideas and delve into core mechanisms of AI and IoT. The ultimate goal of the project is to enhance students' confidence in the concepts and aspects mentioned above and provide them with valuable learning experiences that foster twenty-first-century skills, including creativity, critical thinking, problem-solving, and collaboration.

This project was demonstrated in a small-scale workshop to experienced STEM teachers, who provided preliminary evaluation results on the usefulness of this learning intervention in STEM education and in familiarizing students with the field of AI and IoT. Overall, the teachers expressed strong positivity about the effectiveness of such a project for meaningfully introducing students to AI and IoT. There were concerns about the ability of younger students (under 13) to implement this project, as well as about some activities that are prone to malfunctions and may lead to disappointment. Despite these concerns, many teachers have expressed a high level of interest in gaining access to the produced educational material (teacher guidelines and student worksheets) in order to implement this project with their students. Currently, the project is being piloted with a team of junior high school students. The preliminary feedback so far has been promising. According to the supervising teacher's report, the students are enthusiastic about the results and eager to continue with the rest of the project.

In the near future, the project will also be piloted and evaluated in upcoming teacher training and student pilots, planned within the framework of the AI4STEM project. Based on overall feedback, refinements will be made to the project's content and the included activities to better meet the needs of the target group of the project. Moreover, future steps will involve research toward extending the project by integrating additional AI services, such as image or face recognition, with a view to enhancing automated driving by enabling the recognition of traffic signs or pedestrians. In addition, alternative and age-appropriate technologies and tools will be investigated to assist the production of learning activities that facilitate a smooth introduction of AI and IoT in secondary education.

References

[1]   Chassignol M, Khoroshavin A, Klimova BA. Artificial Intelligence Trends in Education: A Narrative Overview. In: Proc. of the 7th International Young Scientists Conference on Computer Science. Procedia Computer Science; 2018. vol. 136. pp. 16–24.

[2]   Tuomi I. The Impact of Artificial Intelligence on Learning, Teaching, and Education. Policies for the Future. In: Cabrera M, Vuorikari R, Punie Y, eds. EUR 29442 EN. Luxembourg: Publications Office of the European Union; 2018. pp. JRC113226. ISBN 978–92-79-97257-7 doi: 10.2760/12297.

[3]   Casal-Otero L, Catala A, Fernandez-Morante C, Taboada M, Cabreiro B, Barro S. AI literacy in K-12: A systematic literature review. International Journal of STEM Education, Open Access 2023, 10(29), https://doi.org/10.1186/s40594-023-00418-7

[4]   Chiu TKF, Xia Q, Zhou X, Chai CS, Cheng M. Systematic literature review on opportunities, challenges and future research recommendations of artificial intelligence in education. Journal of Computers and Education: Artificial Intelligence, 2023, 4, 100118. https://doi.org/10.1016/j.caeai.2022.100118

[5]   Moroianu N, Iacob SE, Constantin A. Artificial Intelligence in Education: A Systematic Review. In: Proc. of the 6th international conference on economics and social sciences. pp. 2–23. doi: 10.2478/9788367405546-084.

[6]   UNESCO. K-12 AI curricula: A mapping of government-endorsed AI curricula. United Nations Educational, Scientific and Cultural Organization, 2022, https://unesdoc.unesco.org/ark:/48223/pf0000380602 (accessed 13 July 2024).

[7]   Tuomi I. Research for CULT committee – the use of artificial intelligence (AI) in education. May 2020. https://www.europarl.europa.eu/RegData/etudes/BRIE/2020/629222/IPOL_BRI(2020)629222_EN.pdf; (Accessed 13 July 2024).

[8]   Wong G, Ma X, Dillenbourg P, Huan J. Broadening artificial intelligence education in K-12: Where to start?" ACM Inroads, 2020, 11, 20–29. doi: 10.1145/3381884.

[9]   Taguma M, Feron E, Lim MH. Future of Education and Skills 2030: Conceptual Learning Framework Education and AI: Preparing for the Future & AI, Attitudes and Values. In: 8th informal working group (IWG) Meeting. Centre, Paris, France: OECD Conference; October 2018. pp. 29–31.

[10]   Higuera C. A Report about Education, Training Teachers and Learning Artificial Intelligence: Overview of Key Issues. In: Knowledge societies division, communication and information sector of the UNESCO. Universite de Nantes; 2019.

[11]   Touretzky D, Gardener-McCune C, Martin F, Seehorn D. Envisioning AI for K-12: What Should Every Child Know about AI?. In: Proc. of the 33rd AAAI conference on artificial intelligence (AAAI-19), Association of the Advancement of Artificial Intelligence; 2019. pp. 9795–9799.

[12]   Heinze C, Haase J, Higgins H. An Action Research Report from a Multi-year Approach to Teaching Artificial Intelligence at the K-6 Level. In: Proc. of the first symposium of educational advances n AI. Atlanta, Georgia: EAAI; 2010.

[13]   Holmes W, Bialik M, Fadel C. Artificial intelligence in education: Promises and implications for teaching and learning. Boston: Centre for Curriculum Redesign; 2019.

[14]   Kahn K, Megasari R, Piantari E, Junaeti E. AI Programming by Children Using Snap! Block Programming in a Developing Country. In: Proc. of European Conference on Technology-Enhanced Learning. Leeds, UK; 2018.

[15]   Williamson B, Eynon R. Historical threads, missing links and future directions in AI in education. Learning, Media and Technology, 2020, 42(3), 223–235. doi: 10.1080/17439884.2020.1798995.

[16]   Kim J, Lee H, Cho YH. Learning design to support student-AI collaboration: Perspectives of leading teachers for AI in education. Education And Information Technologies, 2022, 1–36. doi: 10.1007/s10639-021-10831-6.

[17]   Gillani N, Eynon R, Chiabaut C, Finkel K. Unpacking the "Black Box" of AI in Education. Educational Technology & Society, 2023, 26(1), 99–111. https://doi.org/10.30191/ETS.202301_26(1).0008

[18]   Papasarantou C, Alimisis D, Geramani K, Ioannidis G, Theodoropoulos E. Artificial Intelligence for Young Students: The Edu4AI Project Handbook. In: Proc. of 14th conference on informatics in education (CIE). Greek Computer Society; 2002. pp. 21–31.

[19]   Papasarantou C, Alimisis D, Geramani K, Ioannidis G. Introducing Artificial Intelligence in School Education: The Edu4AI Project. In: Proc. of the 17th edition of the EUTIC – In the intersection of art, science and technology: Dialogos between humans and machines. Corfu, Greece: Department of Audio & Visual Arts, Ionian University; 2002. pp. 272–280.

[20]   Fleming DS. A teacher's guide to project-based learning. AEL Inc. 2000, available online: https://files.eric.ed.gov/fulltext/ED469734.pdf

[21]   Thomas JW. A review of research on project-based learning. PhD research Supported by the Autodesk foundation, 2000, available online: http://www.bobpearlman.org/BestPractices/PBL_Re search.pdf

[22]   George S, Leroux P. Project-Based Learning as a Basis for a CSCL Environment: An Example in Educational Robotics. In: First European Conference on Computer-Supported Collaborative Learning (Euro-CSCL 2001). Maastricht (Netherland); 2001. pp. 269–276.

[23]   Blikstein P. Digital Fabrication and 'Making' in Education: The Democratization of Invention. In: Walter-Herrmann J, Büching C, eds. FabLabs: of machines, makers, and inventors. Bielefeld: Transcript Publishers; 2013. pp. 203–220.

[24]   Kahn K, Winters N. Constructionism and AI: A History and Possible Features. In: Proc. of the 2020 constructionism conference. The University of Dublin; 2020. pp. 231–236.

[25]   Schon S, Kumar S, Ebner M. The maker movements implications from modern fabrication, new digital gadgets, and hacking for creative learning and teaching. eLearning papers, transforming education through innovation and technology. Special edition. 2014. pp. 86–100.

[26]   Bilkenstein P. Maker Movement in Education: History and Prospects. In: de Vries M, ed. Handbook of technology education. Springer International Handbook of Education; 2018. pp. 419–437.

[27]   Alimisis D. Emerging Pedagogies in Robotics Education: Towards a Paradigm Shift. In: Pons J, ed. Inclusive robotics for a better society, INBOTS 2018, biosystems & biorobotics. Springer; 2020. vol. 25. pp. 123–130. https://doi.org/10.1007/978-3-030-24074-5_22

[28]   Kalogiannakis M, Tzagaraki E, Papadakis S. A systematic review of the use of BBC micro:bit in primary school. New Perspectives in Science Education International Conference, 2021, doi: 10.26352/F318_2384-9509.

[29]   Alimisis D. Technologies for an inclusive robotics education. Open Research Europe, 2021, 1(40), https://doi.org/10.12688/openreseurope.13321.2

List of contributors

Chapter 1
Georgios Lampropoulos
Department of Applied Informatics
School of Information Sciences
University of Macedonia
Greece
And
Department of Education
School of Education
University of Nicosia
Cyprus
And
Department of Preschool Education
School of Education
University of Crete
Greece

Stamatios Papadakis
Department of Preschool Education
School of Education
University of Crete
Greece

Chapter 2
Rina Zviel-Girshin
Ruppin Academic Center
Israel

Nathan Rosenberg
Paralex Institute
Israel

Chapter 3
Ana Carolina Carius
Universidade Católica de Petrópolis
Brazil

Felipe de Oliveira Baldner
Universidade Católica de Petrópolis
Brazil

Chapter 4
Nerea López-Bouzas1
Department of Education Sciences
University of Oviedo
Spain

Jonathan Castañeda Fernández
Department of Education Sciences
University of Oviedo
Spain

M. Esther del Moral Pérez
Department of Education Sciences
University of Oviedo
Spain

Chapter 5
Daniela Conti
Department of Humanities
University of Catania
Catania
Italy

Carla Cirasa
Department of Humanities
University of Catania
Catania
Italy

Santo F. Di Nuovo
Department of Educational Sciences
University of Catania
Catania
Italy

Chapter 6
Rita Cersosimo
DISFOR Department
School of Education
University of Genoa
Genoa

Valentina Pennazio
DISFOR Department
School of Education
University of Genoa
Italy

https://doi.org/10.1515/9783111352695-015

Chapter 7
Bettina Trixler
Doctoral School of Health Sciences
Faculty of Health Sciences
University of Pécs
Hungary

Henriette Pusztafalvi
Department of Health Promotion and Public Health
Faculty of Health Sciences
University of Pécs
Hungary

Chapter 8
Juan-Francisco Álvarez-Herrero
Department of General and Specific Didactics
Faculty of Education
University of Alicante
Spain

Chapter 9
Pedro Baena-Luna
Department of Business Management and Marketing
University of Seville
Seville
Spain

Esther García-Río
Department of Economic Analysis and Political Economy
University of Seville
Seville
Spain

Chapter 10
Hızır Dinler
Department of Early Childhood Education
Kilis 7 Aralık University
Türkiye

Chapter 11
Barbora Stenová
Faculty of Mathematics, Physics, and Informatics
Comenius University Bratislava
Slovakia

Karolína Miková
Faculty of Mathematics, Physics, and Informatics
Comenius University Bratislava
Slovakia

Lucia Budinská
Faculty of Mathematics, Physics, and Informatics
Comenius University Bratislava
Slovakia

Chapter 12
Ela Luria
English and Education Department
Levinsky College of Education & Beit Berl College of Education
Israel

Chapter 13
Svetlana Sharonova
Department of Advertising and business communications
Economics Faculty
RUDN University
Russian Federation

Elena Avdeeva
JSC "Moscow information technologies"
Russian Federation

Chapter 14
Chrissa Papasarantou
European Lab for Educational Technology
EDUMOTIVA
Greece

Dimitris Alimisis
European Lab for Educational Technology
EDUMOTIVA
Greece

Elias Theodoropoulos
European Lab for Educational Technology
EDUMOTIVA
Greece

www.ingramcontent.com/pod-product-compliance
Lightning Source LLC
Chambersburg PA
CBHW061353210326
41598CB00035B/5969